Analysis and Recommendations on the Company-Grade Officer Shortfall in the Reserve Components of the U.S. Army

Catherine H. Augustine, James Hosek, Ian P. Cook, James Coley

Prepared for the Office of the Secretary of Defense

Approved for public release; distribution unlimited

NATIONAL DEFENSE RESEARCH INSTITUTE

The research described in this report was prepared for the Office of the Secretary of Defense (OSD). The research was conducted within the RAND National Defense Research Institute, a federally funded research and development center sponsored by OSD, the Joint Staff, the Unified Combatant Commands, the Navy, the Marine Corps, the defense agencies, and the defense Intelligence Community under Contract W74V8H-06-C-0002.

Library of Congress Control Number: 2011927724

ISBN: 978-0-8330-5185-1

The RAND Corporation is a nonprofit institution that helps improve policy and decisionmaking through research and analysis. RAND's publications do not necessarily reflect the opinions of its research clients and sponsors.

RAND® is a registered trademark.

Cover photo by Sgt. John Crosby, U.S. Army

Published 2011 by the RAND Corporation
1776 Main Street, P.O. Box 2138, Santa Monica, CA 90407-2138
1200 South Hayes Street, Arlington, VA 22202-5050
4570 Fifth Avenue, Suite 600, Pittsburgh, PA 15213-2665
RAND URL: http://www.rand.org/
To order RAND documents or to obtain additional information, contact
Distribution Services: Telephone: (310) 451-7002;
Fax: (310) 451-6915; Email: order@rand.org

Preface

A June 18, 2009, U.S. House of Representatives Armed Services Committee report on Public Law 111-84 of the National Defense Authorization Act (NDAA) for Fiscal Year 2010 (House Report [H.R.] 111-166) requested the Secretary of Defense to conduct a study of the company-grade officer shortfall in both Reserve Components (RCs) of the U.S. Army: the U.S. Army Reserve (USAR) and the Army National Guard (ARNG). In response to that report, the Office of the Assistant Secretary of Defense for Reserve Affairs (OASD/RA) asked the RAND National Defense Research Institute (NDRI) to conduct a study on the company-grade officer shortfall in both Reserve Components. This monograph is intended to satisfy this request. As such, it addresses the shortfall of company-grade officers in the USAR and the ARNG. However, our recommendations for the U.S. Army could be applicable to the U.S. Marine Corps (USMC) and the U.S. Navy, which are also experiencing a shortfall in company-grade officers in their Reserve Components.

This monograph addresses four study goals. It explores and validates the magnitude of the company-grade officer shortfall in the Reserve Components of the U.S. Army, concluding that the problem is really a captain shortfall, and identifies recommendations to address this shortfall. In making recommendations, the monograph assesses whether the concept of a National Guard academy is a feasible partial solution to the captain shortfall. It also assesses the impact of requiring Officer Candidate School (OCS) candidates to hold a four-year degree to participate in OCS. The study was conducted from October 2009 through March 2011.

This research was sponsored by OASD/RA and conducted within the Forces and Resources Policy Center of the RAND National Defense Research Institute, a federally funded research and development center sponsored by the Office of the Secretary of Defense, the Joint Staff, the Unified Combatant Commands, the Navy, the Marine Corps, the defense agencies, and the defense Intelligence Community.

For more information on the RAND Forces and Resources Policy Center, see http://www.rand.org/nsrd/ndri/centers/frp.html or contact the director (contact information is provided on the web page).

Contents

Figures

Tables

Summary

For more than a decade, the U.S. Army National Guard (ARNG) and the U.S. Army Reserve (USAR) have both had lower inventories of company-grade officers than have been authorized.[1] Company-grade officers include both lieutenants and captains. The shortage is in the rank of captain: For more than five years now, both Army Reserve Components (RCs) have had higher inventories of lieutenants than authorized but lower inventories of captains than authorized. Therefore, this monograph focuses on the captain shortfall.

Current literature on the captain shortfall in the RCs posits that the initial cause of the shortfall lies in the reduction of commissions in the early 1990s coinciding with the military drawdown, and the struggles experienced with the early implementation of the Reserve Officer Personnel Management Act (ROPMA, part of Pub. L. 103-337), which resulted in an unnecessary loss of officers. Further compounding the problem, the Reserve Officer Training Corps (ROTC) failed to meet mission requirements for nearly a decade and a half, necessitating recruiting from the potential RC officer pool to adequately staff the active force. Attrition is not a primary driver of the shortfall. Although several studies have expressed concern about retaining captains in the RCs due to high rates of deployments, data demonstrate that captain loss rates have remained steady to improving over the past ten years. Past studies point to the continued deficit as stemming from multiple causes, a combination of factors that have sustained, if not increased,

---

[1] Congress sets authorizations for total end-strength objectives (per Title 10 of the U.S. Code, Chapter 1201); the Army then sets authorizations by rank.

the decade-old problem of insufficient captains to meet RC authorizations. This reveals the complexity of the issue and suggests the absence of a single remedy for correcting the shortfall.

A June 18, 2009, House Armed Services Committee report on Public Law 111-84 of the National Defense Authorization Act (NDAA) for Fiscal Year 2010 (House Report [H.R.] 111-166) requested the Secretary of Defense to conduct a study on the company-grade officer—and, in particular, captain—shortfall in the RCs of the U.S. Army. Although that report was not included in the final NDAA, the Office of the Assistant Secretary of Defense for Reserve Affairs (OASD/RA) asked the RAND National Defense Research Institute (NDRI) to conduct a study on the company-grade officer shortfall in the USAR and the ARNG. Page 314 of H.R. 111-166 for Public Law 111-84 provides the motivation for this study:

> The committee understands that the Army National Guard and Army Reserve have historically been challenged with company grade officer shortages, primarily at the captain (0–3) rank. The reasons for these shortages stem from a number of issues, including the difficulty officers have in meeting the requirement for a bachelor's degree as a condition for promotion to captain.
>
> The committee is concerned that this shortage of company grade officers needs to be addressed if the Army National Guard and Army Reserve are to be an effective part of the operational reserve force. Therefore, the committee directs the Secretary of Defense, in consultation with the Chief of the National Guard Bureau and the Chief of the Army Reserve, to conduct a comprehensive study of this issue and to make recommendations on how to address these officer shortages. The study should include:
>
> (1) A review of the concept of a National Guard military academy, similar to the service academies including the following:
>
> whether such a National Guard academy is a feasible partial solution to the officer shortages and, if feasible, the roles and responsibilities for operating a military academy; the estimated costs for the establishment of an academy; the annual operating costs, to

include staffing requirements and academic faculty requirements to meet accreditation requirements of a four-year institution of higher learning; and the ability to incorporate junior military colleges into the program. It should also address: issues of compulsory service obligations; the challenges involved with granting commissions to cadets from different states; how funding for students and resources for the academy might be provided; what academic programs the academy might offer; the admissions process; the training requirements for cadets/students; and the number of cadets/students that would have to be authorized each school year.

(2) A consideration of the feasibility of requiring state Officer Candidate School [OCS] programs to require candidates to hold a four-year degree in order to participate in the program, and the necessary programmatic changes that may be required to support such a requirement.

The committee directs the Secretary to report his findings, conclusions, and recommendations to the Senate Committee on Armed Services and the House Committee on Armed Services within one year after the date of enactment of this Act.

This monograph is intended to satisfy this request. As such, it addresses the shortfall of company-grade officers in general, and captains in particular, in the USAR and the ARNG. However, our recommendations for the USAR could be applicable to the U.S. Marine Corps (USMC) Reserve and the U.S. Navy Reserve, which are also experiencing a shortfall in company-grade officers in their RCs.

This monograph has four intents. It explores and confirms the magnitude of the company-grade and captain-only shortfall in the RCs of the U.S. Army. It identifies recommendations to address the captain shortfall. In making recommendations, the monograph assesses whether the concept of a National Guard academy is a feasible partial solution to the company-grade officer shortfall. It also assesses the impact of requiring OCS candidates to hold a four-year degree to participate in OCS.

The study relies on analysis of Structure and Manpower Allocation System (SAMAS) and Defense Manpower Data Center (DMDC) data through fiscal year (FY) 2009. SAMAS and DMDC data were used to confirm the magnitude of the shortfall and to conduct inventory modeling to project future fill rates for captains in both RCs. We considered the total officer populations in both RCs, including full-time Active Guard and Reservists, individual mobilization augmentees, and those in special branches (e.g., chaplains, lawyers, and medical professionals).

To generate recommendations for addressing the shortfall, we relied on our internal data analysis and on ideas emanating from 25 interviews that we conducted with more than 50 participants. Participants included representatives from the Department of the Army; Army component manpower or personnel staff officers (G-1) and subordinate commands; the National Guard Bureau; the ARNG; five adjutant generals (TAGs) and their staffs; the Office of the Chief of the Army Reserve (OCAR); Army Reserve, G-1; U.S. Army Reserve Command (USARC); and Army Reserve Careers Division (ARCD). We also interviewed the USMC Reserve Affairs staff about their company-grade officer shortfall.

Fill Rates

Although, in both Army RCs, the captain inventory-to-authorization fill rate has fallen short of 80 percent for the past seven years, the overall captain and company-grade fill rates have been improving. In the ARNG, the lieutenant fill rate has exceeded authorizations since 2004, and the captain fill rate has improved since 2006. The overall ARNG company-grade fill rate is 95 percent, with a 71-percent captain fill rate. These rates have been increasing since 2006, primarily as a result of an increasing inventory. In the USAR, the lieutenant fill rate has exceeded authorizations since at least 2003, and the captain fill rate has improved since 2005; the FY 2009 fill rate was 75 percent. The FY 2009 overall USAR company-grade fill rate was 98 percent. This rate has been increasing since 2007 because of an increasing inventory, as well as decreasing authorizations.

Although this monograph focuses on company-grade officers, we observed that there is also a shortfall in the inventory of majors, compared to authorizations, in both Army RCs. In FY 2009, the major fill rate in the ARNG was 74 percent; it was 89 percent in the USAR. The gap between major authorizations and inventory appears to be long-standing, particularly in the ARNG. In the USAR, the major fill rate has declined by about 30 percentage points over seven years. The number of majors in each RC has decreased over the past seven years while the authorizations have increased in the ARNG and remained fairly stable in the USAR. The major shortfall is worsening as the captain shortfall is improving, albeit moderately.

Inventory Projection Modeling

Our inventory projection modeling indicates that the Army RCs could achieve a 100-percent captain fill rate in five to ten years, assuming accelerated promotion rates and positive responses to incentives and therefore favorable accession and continuation rates. The captain fill rate could increase to 90 percent fairly soon—in about two to four years for the ARNG and in one to three years for the USAR, if both RCs can increase accessions and promotion rates while maintaining recent loss rates.

We tested ten different models to generate these results. We assumed both exponential increases in accessions and a large increase over the next three years, followed by stabilization. We tested a gain of 4 percent (in the ARNG) and 6 percent (in the USAR) more company-grade officers each year in the exponential growth model. These gains are based on high, but demonstrable, year-to-year gains observed in the data over the past ten years. These gains include the numbers of officers transferring from the Active Component (AC), as well as those starting their officer career in an RC. In the ramp-up model, we assumed large growth in accessions over the next three years (10, 20, and then

30 percent more than was observed in 2009), followed by no growth in future years.

The modeling assumes that the average officer loss rates that we observe in both RCs over the past ten years continue, which is 8.61 percent for the ARNG and 10.39 percent for the USAR. Although loss rates have been declining from peaks in 2004–2005 (for ARNG) and 2005 (for USAR) on the order of 2 percentage points for ARNG and 4 percentage points for USAR each year, we base our projections on ten-year historical averages rather than on the recent higher rates.

The modeling also tests promoting lieutenants to captain more quickly than is the norm[2] and keeping captains in grade for longer than is the norm.[3] Although the latter might be technically feasible, it could have unintended negative consequences on retention rates.

The most-aggressive model results in a 100-percent captain fill rate in the ARNG is 6.5 years and five years in the USAR. In this model, we assume that accessions increase by 10 to 30 percent over the next three years, followed by a leveling-off period. This modeling assumes historical average continuation and promotion rates (if lieutenants were promoted to captain more quickly, the captain fill rate could reach 100 percent even earlier). Accessioning 10 percent more officers in 2010, followed by 20 percent more than the 2009 numbers the following year and 30 percent more the next, would entail aggressive accessioning policies and practices, but it could be worth investing in such policies and practices in the short term, particularly given the current unemployment context, to dramatically boost the captain fill rate.

---

[2] Data on second lieutenants newly commissioned in 2002 indicate that, seven years later, by 2009, 56 percent of these ARNG second lieutenants and 63 percent of these USAR second lieutenants had been promoted to captain. In the ARNG, of those who were promoted in seven years, most (56 percent) were promoted with five to six years time in grade (TIG) as a lieutenant, but a significant proportion (38 percent) was promoted with between six and seven years TIG as a lieutenant. In the USAR, of those who were promoted in seven years, about half (48 percent) were promoted with five to six years TIG as a lieutenant, and 46 percent were promoted with between six and seven years TIG as a lieutenant.

[3] Across the two RCs, the 2002 new captain cohort remained in captain grade for an average of five years.

The most-conservative model results in a 100-percent captain fill rate in 10.5 years in the ARNG and nine years in the USAR. In this model, accessions increase by 4 percent in the ARNG and 6 percent in the USAR each year for the next 10.5 years. Recent average loss rates continue. Lieutenants are promoted to captain after serving as a second and then a first lieutenant for a total of four years. On average, captains are promoted to major after five years, which is the current average TIG.

Problems Associated with the Shortfall

Both the literature and the fill-rate data suggest that readiness could be compromised by a shortage of captains. We attempted to verify this hypothesis through primary and secondary data analysis. We explored whether the following problems exist in the ARNG or USAR as a result of an insufficient number of captains:

- Units are less likely to be deemed deployable.
- There is more cross-leveling due to shortfall.
- Promotion to major is slowed.
- Lieutenants have insufficient time to develop before they are asked to lead units.
- Captains are deployed more frequently than are other officers.

We also explored the extent to which promotion to major has been slowed and found that it has not slowed and that, on average, captains are not spending the maximum allowable time in grade. Although the literature also speculates that through "fully qualified"[4] promotion

---

[4] Guided by Army Regulation (AR) 600-8-29, Officer Promotions, *fully qualified* means qualified "professionally and morally to perform the duties expected of an officer in the next higher grade" (p. 19). *Best qualified* means fully qualified officers who "meet specific branch, functional area or skill requirements" (p. 19).

Promotion boards will do the following:

(3)(a) The "fully qualified" method when the maximum number of officers to be selected, as established by the Secretary, equals the number of officers above, in, and below the

practices, the quality of company-grade officers could be declining, we did not attempt to verify this hypothesis.

Interviewees disagreed that units are less likely to be deemed deployable but did acknowledge that unit readiness was achieved by cross-leveling. Interview data were inconclusive on the extent to which cross-leveling occurs because of the shortfall in captains. Certainly, cross-leveling happens in response to personnel readiness needs, but those needs include not only rank needs but also military occupational specialty (MOS)–qualified/senior staff needs.

Most vacant captain slots are filled by first lieutenants (27 percent), followed by second lieutenants (9 percent) and majors (9 percent). Interviewees disagreed that lieutenants are put into leadership positions with insufficient training and experience. When interviewed, RC senior leaders and TAGs both stressed that, given the excess of lieutenants, they are able to select qualified ones for captain slots.

Finally, we did not observe higher deployment rates for captains than for lieutenants and majors.

Data analyses and responses to our interview questions on the severity of the captain shortfall led us to question the extent to which the shortfall is problematic. Interviewees struggled to describe problems caused by the captain shortfall. It could be that, because the shortfall has been in place for several years, commanders have become accustomed to finding alternative methods to solve the issue and no longer view it as a significant problem. If that is the case, it could be

---

promotion zone. Although the law requires that officers recommended for promotion be "best qualified" for promotion when the number to be recommended equals the number to be considered, an officer who is fully qualified for promotion is also best qualified for promotion. Under this method, a fully qualified officer is one of demonstrated integrity, who has shown that he or she is qualified professionally and morally to perform the duties expected of an officer in the next higher grade. The term "qualified professionally" means meeting the requirements in a specific branch, functional area, or skill.

(3)(b) The "best qualified" method when the board must recommend fewer than the total number of officers to be considered for promotion. However, no officer will be recommended under this method unless a majority of the board determines that he or she is fully qualified for promotion. As specified in the MOI [memorandum of instruction] for the applicable board, officers will be recommended for promotion to meet specific branch, functional area or skill requirements if fully qualified for promotion. (p. 19)

worth investigating the extent to which the current authorizations for captains are warranted. However, it could also be true that this short-fall is indeed highly problematic and that we would have learned more about the problems had we interviewed commanders in the field.

Policy Options

Even if the current authorization structure is modified, demands for captains (and majors) are likely to outpace supply in the next several years. We therefore present policy options for increasing accessions and improving promotion and retention rates. The current unemployment context should facilitate efforts to improve both accessions and retention rates in the near term. In considering accessions, we focus on AC-to-RC affiliations and on the three largest sources of new commissions in the RCs: ROTC, the Officer Direct Commission (ODC) program, and the state OCS programs.

Active Component–to–Reserve Component Transfers

According to 10 U.S.C. 651, AC officers who have a remaining military service obligation (MSO) and are otherwise qualified, shall, upon release from active duty, be transferred to an RC of his or her armed force to complete the service required. The Army estimates that the RCs are gaining approximately 40 percent of officers who leave the AC with ten or fewer years in service and are eligible for an RC commission. Human Resource Command (HRC) interviewees also reported several barriers to this transfer process. Because these barriers appear to be surmountable, and because gaining more transfers results in an immediate boost in the number of captains, it could be worth investing in increasing this rate. Earlier counseling on RC options, combined with options to improve the process of moving from the AC to an RC, could improve this affiliation rate.

Moving to a seamless interservice transfer process could necessitate legislative and regulatory changes. One option to improve the transfer process is to change the statute to allow a single commission. Another option would be to allow an exiting AC officer to sign the RC commission at the time of exit and wait to date the RC commission until the AC resignation is approved. A third option is to offer exiting AC officers the opportunity to sign a letter of intent to join an RC and offer an increased bonus (perhaps selectively, depending on the extent of the shortage and the criticality of the officer's specialty) as an incentive to sign. Other incentives might include graduate school reimbursement or jobs with federal agencies.

Reserve Officer Training Corps

A recent ROTC policy change should benefit RC recruiting, although more could be done to incentivize ROTC cadets to join the RCs. The policy change requires that each component (i.e., Army AC, ARNG, and USAR) achieve its mission before it can overrecruit. This new policy should ensure that the RCs get their "fair share" of the ROTC cadets. However, both Army RCs would prefer to incentivize cadets into an RC than commission those who would prefer to join the AC but score at the bottom of the order-of-merit list (OML). At the point of contracting, cadets need information about the RCs so that they can make an informed choice about components and the scholarships each can offer. Cadets could also be offered a range of incentives to join an RC, including any combination of the following:

- bonus
- choice of branch (with caveats related to regional variation)
- guaranteed internships
- simultaneous eligibility for Dedicated Army National Guard (DEDNG) scholarship students for the Montgomery GI Bill–Selected Reserve (MGIB-SR, or 10 U.S.C. Chapter 1606)
- graduate school funding.

Many interviewees support the notion of changing ROTC and U.S. Military Academy (USMA) contracts so that those who join the AC are required, at the service's discretion, to fulfill their remaining MSO in a drilling unit rather than in the Individual Ready Reserve (IRR).

The Officer Direct Commission Program

There is a proposal under consideration to directly commission more captains into the USAR and ARNG. There are concerns, however, about the ability of newly commissioned captains to lead troops. The proposal limits troop-leading responsibilities in the early phases of these new captains' careers. Indeed, it should be possible to utilize direct commissions more extensively for occupations that draw on skills acquired as a civilian for initial assignments within staff organizations, accompanied by expanded leadership education and training to prepare these officers for subsequent assignments within operational units.

Army National Guard State Officer Candidate School Programs

H.R. 111-166 requests the Secretary of Defense to address whether soldiers entering state OCS programs should possess a baccalaureate degree. Currently, officers can be commissioned as lieutenants without a baccalaureate degree, but they must obtain that degree for promotion to captain.[5] The motivation for this question stems from the concern that many officers who graduate from a state OCS program never complete their degree and, therefore, are never promoted to captain. Although 42 percent of the lieutenants commissioned via OCS without a four-year degree in 2001 had left the service by 2008,[6] 58 percent were promoted to captain (meaning that they had obtained their degree) and were still serving in 2008. This proportion (58 percent) is less than the 72 percent of those who entered OCS with a four-year degree and were promoted to captain and were still serving in 2008.

Despite this statistically significant difference, we recommend continuing the practice of allowing soldiers with 60 or more credit hours to start an OCS program. Policy decisions need to be made within a context of a shortfall in the ARNG captain inventory. Most of the OCS soldiers who lacked four-year degrees in 2001 did become

---

[5] Lieutenants also must have completed Basic Officer Leaders Course (BOLC B), a military education branch and basic soldiering course, before they are eligible for promotion to captain.

[6] It is possible that some of these officers had obtained a four-year degree prior to leaving.

captains. Furthermore, more than one-third of the 2001 cohort who made it to captain and were still in service in 2008 started without a four-year degree.

We also recommend, however, that OCS candidates complete their four-year degree while they are going through OCS and that such completion be required for a commission into an RC. We believe, based on the literature and our interview data, that a soldier would have more time and support during the traditional OCS programs to complete a degree than he or she would postcommissioning, when lieutenants are frequently deployed. The Army could choose, however, to defer implementing this policy until the captain fill rate has sufficiently improved.

National Guard Academy

H.R. 111-166 also requested the Secretary of Defense to consider whether establishing a new National Guard academy would be a partial and feasible solution to the company-grade officer shortfall. To address this question, we relied on projections published by Danner (2010) that a new National Guard academy could enroll 250 cadets annually, starting in 2015. We assumed that all of these cadets would graduate in 2019 and be promoted to captain by 2023. By then, assuming that officers respond to new incentives, our inventory projection modeling indicates that the ARNG could have met a 100-percent captain fill rate through aggressive accessioning and promotion practices and the maintenance of recent average continuation rates.

Although these accession practices are likely to have costs associated with them, these costs would unlikely approximate those of establishing and continuing to operate a new postsecondary institution. Although we are not certain of what the average cost per graduate from a new National Guard academy would be, if the caliber of the academy were like that of West Point, Annapolis, and the Air Force Academy, a cost of $400,000 per graduate would not be out of line. Therefore, although we acknowledge that a new academy is a feasible source to commission new officers, it would not eliminate the shortage any sooner than would other methods and would likely cost more. Consequently, we conclude that establishing a new National Guard academy is not a cost-effective solution to the company-grade officer shortfall. More cost-effective options include the AC-to-RC transfer, ROTC, OCS (including the OCS enlistment option), federal OCS (including those without prior service), and modifying ROTC contracts to mandate service in the Selected Reserve (SELRES), rather than allowing the completion of one's MSO in the IRR.

We acknowledge, however, that establishing such an academy could have other educational and leadership benefits that we have not assessed in this monograph. It might address concerns in the ARNG other than the company-grade officer shortfall and could, therefore, be an effective solution to such problems.

Promotion

We recommend that both Army RCs promote first lieutenants to captain more quickly than has been the case. In addition to modestly increasing accessions, promoting an officer from a second lieutenant to a first lieutenant and then to captain in four years, on average, would allow both RCs to reach a 100-percent captain fill rate in nine to ten years. The TIGs necessary to promote a lieutenant to captain in four years fall within statutory limits. But promoting someone to captain with four years time in service (TIS) as an officer represents a change in practice: Ninety-four percent of the officers who started as a lieutenant in 2002 and were promoted within seven years had spent at least five years in total as a second and first lieutenant.

Retention

Because the literature, our interviewees, and our data analyses all support the conclusion that current retention rates are higher than historical rates, we did not manipulate loss rates in our modeling. However, the ARNG and USAR could benefit from improved retention rates and might also want to strategize now on how to sustain high retention rates as the economy improves. Incentives, such as bonuses or support for graduate education, could be effective policy tools for increasing or maintaining current retention rates.

Further Study

In this monograph, we have focused on increasing the supply of captains; we also recommend that the Army conduct an analysis of its force structure. Key questions that could be addressed in such an analysis include the following:

1. Which captain positions are being filled by lieutenants or are being left vacant?
2. Could these positions be recoded to a lower rank?
3. For those positions filled by lieutenants, is there a cascading effect that leaves lieutenant positions vacant?
4. What is the impact of cross-leveling officers across positions requiring different ranks?

Understanding the specific requirements of the vacant positions could provide guidance on reclassifying positions, direct commissioning at higher ranks, or even eliminating the authorization.

We suggest in this monograph that there could be room for increasing the AC-to-RC transfer rate. Further analysis could explore AC officers' behavior and motivation and build demonstration projects that address them. Data from interviews with a sample including both exiting and remaining officers could be analyzed to better understand those officers' decisionmaking processes. These data could be used to develop approaches (e.g., better counseling, higher incentives) to increase the proportion that transfers to an RC.

In terms of promotion practices, further analysis could facilitate the ability of the RCs to promote lieutenants to captain at minimum TIG and accelerate the use of vacancy boards, particularly in the USAR. It would be important to understand the barriers to vacancy promotions, as well as the limits of this system in general. In general, major fill rates should be addressed.

Although this study concluded that a National Guard academy is not a cost-effective solution to the company-grade officer shortfall, it could be worth studying whether and how ARNG officer education could be improved. Further study could focus on curriculum design, investigating the content of current curricula, and ascertaining whether additional or different emphases could benefit the dual mission of the ARNG. The current proposal to establish a new academy has reportedly generated excitement among several constituencies, which could indicate a need for new ARNG education models.

Conclusion

This monograph is a response to the June 18, 2009, House Armed Services Committee report (H.R. 111-166) that requested that the Secretary of Defense conduct a study on the company-grade officer—and, in particular, captain—shortfall in the RCs of the U.S. Army. We found that, although the overall company-grade officer fill rates in Army RC units are improving gradually due to slowly increasing captain fill rates and lieutenant fill rates that increasingly exceed 100 percent, aggressive measures would be needed to dramatically improve the captain fill rate, and thus the overall company-grade officer fill rate, in both RCs. Our modeling demonstrates that the Army RCs could achieve a 100-percent captain fill rate in five to ten years if they can sustain recent low loss rates, increase officer accession rates, and promote lieutenants to captain more quickly (but within statutory TIG limits). This modeling indicates that the ARNG could achieve a 100-percent captain fill rate before a new National Guard academy would have produced captains. Although our modeling assumes increased accessioning, which will necessitate resources, these costs would not likely approximate those of establishing and continuing to operate a new postsecondary institution. Therefore, although we acknowledge that a new academy is a feasible source of new captains, it would not eliminate the shortage any sooner than would other methods and would very likely cost more.

We do recommend that the Army conduct an analysis of its force structure with a specific focus not only on captains but on the rank of major as well. Our analysis indicates that the captain shortfall is migrating up to the rank of major. Understanding the specific requirements of the vacant captain and major positions could provide additional guidance on reclassifying positions, direct commissioning at higher ranks, or even eliminating the authorization. We also recommend several strategies to increase accessioning. Specifically, we recommend increasing the number of transfers from the AC into the RC, incentivizing ROTC cadets to join an RC early in their educational tenure, and increasing the number of captains directly commissioned. We do not recommend requiring state ARNG OCS entrants to hold a baccalaureate degree.

Acknowledgments

Several people contributed to this study and monograph. We would like to thank OASD/RA for substantive assistance. Staff from this office provided valuable guidance on the analytic components of our study. Additionally, several personnel from the U.S. Army G-1 were instrumental in coordinating interviews, fielding data and other questions, and providing suggestions to improve our work.

We also want to thank the individuals we interviewed for this study. We are particularly grateful for the time given to us by these interviewees who represented the Department of the Army (DA), ARNG, USAR, and USMC. Although we are keeping their identities confidential, their knowledge, suggestions, and expertise contributed greatly to this study.

COL Jeffrey D. Peterson, Academy Professor of Economics at West Point Military Academy, and Ellen Pint, a senior economist at RAND, reviewed an earlier draft of this monograph, and their insightful comments greatly improved it. Colonel Peterson provided suggestions for conducting a force-structure analysis that we have adopted in this monograph.

Within RAND, several people contributed to this work. John Winkler provided guidance and feedback. Laurie McDonald extracted and organized the necessary data. Ian Cook conducted the inventory projection modeling. Stephanie Lonsinger provided research and editing support.

Abbreviations

1LT	Army first lieutenant
2LT	Army second lieutenant
AC	Active Component
ADSO	active-duty service obliation
AGR	Active Guard and Reserve
APL	Army Promotion List
AR	Army regulation
ARCD	Army Reserve Careers Division
ARFORGEN	Army Force Generation
ARNG	Army National Guard
ASA(M&RA)	Assistant Secretary of the Army for Manpower and Reserve Affairs
BOLC	Basic Officer Leaders Course
BrADSO	branch-of-choice for active-duty service obligation
CC	Cadet Command
CGSOC-RC	Command and General Staff Officers' Course, Reserve Component

CPT	Army captain
DA	Department of the Army
DEDNG	Dedicated Army National Guard
DEERS	Defense Enrollment Eligibility Reporting System
DMDC	Defense Manpower Data Center
E1, E2	pay grades for Army privates
E3	pay grade for Army private first class
E4	pay grade for Army corporal or specialist
E5	pay grade for Army sergeant
E6	pay grade for Army staff sergeant
E7	pay grade for Army sergeant first class
E8	pay grade for Army master sergeant or first sergeant
E9	pay grade for Army sergeant major or command sergeant major
ECP	Early Commissioning Program
FY	fiscal year
G-1	component manpower or personnel staff officer
GrADSO	Graduate School for Active Duty Service Obligation
GRFD	Guaranteed Reserve Forces Duty
GWOT	global war on terrorism
H.R.	House report
HRC	Human Resource Command
IMA	individual mobilization augmentee

IRR	Individual Ready Reserve
LT	Army lieutenant
LTC	Army lieutenant colonel
MAJ	Army major
MFORCE	Master Force
MGIB-SR	Montgomery GI Bill–Selected Reserve
MJC	military junior college
MOS	military occupational specialty
MSO	military service obligation
NDAA	National Defense Authorization Act
NDRI	National Defense Research Institute
NGB	National Guard Bureau
O1	pay grade for Army second lieutenant
O2	pay grade for Army first lieutenant
O3	pay grade for Army captain
O4	pay grade for Army major
O5	pay grade for Army lieutenant colonel
O6	pay grade for Army colonel
OASD/RA	Office of the Assistant Secretary of Defense for Reserve Affairs
OCAR	Office of the Chief of the Army Reserve
OCS	Officer Candidate School
ODC	Officer Direct Commission
OML	order-of-merit list

OPTEMPO	operating tempo
PERSCOM	Personnel Command
PVB	promotion vacancy board
RC	Reserve Component
RCCPDS	Reserve Component common personnel data system
ROPMA	Reserve Officer Personnel Management Act
ROTC	Reserve Officer Training Corps
SAMAS	Structure and Manpower Allocation System
SELRES	Selected Reserve
SRC	standard requirement code
TAG	[the] adjutant general
TAPDB-R	Total Army Personnel Data Base–Reserve
TIG	time in grade
TIS	time in service
TPU	troop program unit
TRADOC	U.S. Army Training and Doctrine Command
USAR	U.S. Army Reserve
USARC	U.S. Army Reserve Command
USMC	U.S. Marine Corps
UVB	unit vacancy board
WEX	work experience

CHAPTER ONE

Introduction

The Reserve Components (RCs) of the U.S. Army have had less than 100 percent of Army-authorized captains for at least the past ten years. The House Armed Services Committee's House Report (H.R.) 111-166 on the National Defense Authorization Act (NDAA) for Fiscal Year (FY) 2010 (Pub. L. 111-84), published on June 18, 2009, addressed this shortfall. Although H.R. 111-166 language was not included in the FY 2010 NDAA, which was enacted on October 28, 2009, the Secretary of Defense determined that it was necessary to address the shortfall. Page 314 of H.R. 111-166 in Public Law 111-84 provides the motivation for this study:

> The committee understands that the Army National Guard [ARNG] and Army Reserve [USAR] have historically been challenged with company grade officer shortages, primarily at the captain (0–3) rank. The reasons for these shortages stem from a number of issues, including the difficulty officers have in meeting the requirement for a bachelor's degree as a condition for promotion to captain.

> The committee is concerned that this shortage of company grade officers needs to be addressed if the Army National Guard and Army Reserve are to be an effective part of the operational reserve force. Therefore, the committee directs the Secretary of Defense, in consultation with the Chief of the National Guard Bureau and the Chief of the Army Reserve, to conduct a comprehensive study of this issue and to make recommendations on how to address these officer shortages. The study should include:

(1) A review of the concept of a National Guard military academy, similar to the service academies including the following:

whether such a National Guard academy is a feasible partial solution to the officer shortages and, if feasible, the roles and responsibilities for operating a military academy; the estimated costs for the establishment of an academy; the annual operating costs, to include staffing requirements and academic faculty requirements to meet accreditation requirements of a four-year institution of higher learning; and the ability to incorporate junior military colleges into the program. It should also address: issues of compulsory service obligations; the challenges involved with granting commissions to cadets from different states; how funding for students and resources for the academy might be provided; what academic programs the academy might offer; the admissions process; the training requirements for cadets/student; and the number of cadets/students that would have to be authorized each school year.

(2) A consideration of the feasibility of requiring state Officer Candidate School [OCS] programs to require candidates to hold a four-year degree in order to participate in the program, and the necessary programmatic changes that may be required to support such a requirement.

The committee directs the Secretary to report his findings, conclusions, and recommendations to the Senate Committee on Armed Services and the House Committee on Armed Services within one year after the date of enactment of this Act.

In response to that report, the Office of the Assistant Secretary of Defense for Reserve Affairs (OASD/RA) asked RAND to conduct a study on the company-grade officer shortfall in the USAR and the ARNG. This monograph is intended to satisfy that request. As such, it addresses the shortfall of company-grade officers, and captains in particular, in the USAR and the ARNG. However, our recommendations for the U.S. Army could apply to the U.S. Marine Corps (USMC) and the U.S. Navy, which are also experiencing shortfalls in company-grade officers in their RCs.

Background

According to the literature on the subject, which we describe in greater detail later, the company-grade officer shortfall in the Army RCs dates back to the early 1990s. Over the past decade, several studies have addressed the causes of the shortfall, resulting problems, and recommendations to remedy it. Although these past studies have varying degrees of commonality and divergence, all point to the continued deficit as stemming from multiple causes, a combination of factors that have sustained the decade-old problem of insufficient captains to meet RC authorizations. Past research reveals the complexity of the issue and suggests the absence of a single remedy for correcting the shortfall.

Causes of the Shortfall

As noted in H.R. 111-166, the company-grade officer shortfall is really a captain shortfall. To some extent, particularly in the ARNG, the recent conversion to brigade combat team maneuver units dictated the need for more captains (Reno, 2008). However, the captain shortfall exists in both RCs and is at least a ten-year-old problem.

Across the literature, the officer shortfall is described as rooted in a reduction in accessions (Whitlock, 2002; Howe, 2005; Feidler, 2008; Reno, 2008). Attrition rates have been fairly steady through 2009, and Feidler (2008), for example, found no indication that officers were attempting to leave the military in large numbers.

Low accession rates were a result of the active Army drawdown of the 1990s and the concomitant reduction in commissions from the Reserve Officer Training Corps (ROTC). The decline in ROTC production is attributed to the Army's response to the U.S. General Accounting Office's (GAO's; now the U.S. Government Accountability Office) report, Reserve Officers' Training Corps: Less Need for Officers Provides Opportunity for Significant Savings (GAO, 1991). That report was responding to data from 1989. The data suggested not only that had ROTC met its mission but also that there was an over-supply of lieutenants for both the Army and Air Force. However, by the time the report was published, ROTC had failed to meet its mission for two consecutive years and would continue to do so for almost a decade and a half (Howe, 2005).

With ROTC having failed to meet its mission after 1989, fewer lieutenants than needed were joining the Active Component (AC). To meet its authorizations, the active Army was therefore forced to draw on ROTC accessions intended for the RCs. As a result, the reduced 1992–2005 year-groups led to reduced midgrade officer populations in the Army RCs (Howe, 2005; Erlandson, 2009).

To remedy the shortfall, the literature recommends greater accessions through ROTC and other commissioning sources, as well as greater transfers from the AC into the RCs. Because the shortfall is most acute at the captain rank, Feidler (2008) asserted that the USAR must recruit greater numbers of officers from the AC. He argued that many Army officers leaving active duty are entering the Individual Ready Reserve (IRR) instead of filling slots in the Selected Reserve (SELRES) in the USAR and ARNG, a problem that both Reno (2008) and Erlandson (2009) also noted. Furthermore, Feidler asserted that many USAR officers are transferring into the AC, or even to the ARNG, although he did not cite data supporting this observation. And Reno noted that, although the ARNG had achieved 102 percent of authorizations for lieutenants, it would be required to reach 125-percent strength over the next decade to promote itself out of the problem if it did not receive more AC transfers into the SELRES.

Study authors also expressed concern about the promotion system and the role this system has played in creating the current shortfall. The 1996 Reserve Officer Personnel Management Act (ROPMA, part of Pub. L. 103-337) was intended to update and consolidate the laws governing all RC officers to effectively manage the officer career path from appointment to separation. A key provision of the act is the man-datory selection criteria for promotion consideration to captain, which include the requirement of a bachelor's degree, as mandated by the 1995 NDAA (Whitlock, 2002; Bonn, 2005).

Whitlock (2002) argued that the requirement of a bachelor's degree contributed to a loss of qualified officers. He contended that many of the USAR lieutenants who came before promotion boards, though possessing the required baccalaureate degree, nonetheless had no proof of their degree on record when coming before the board and were denied promotion. Whitlock correlated this with a significant drop in officer promotions after 1996, concluding that the percentage of officers who were educationally qualified for promotion dropped in a single year (between 1996 and 1997) from 91 percent in troop program units (TPUs) and 88 percent in the IRR to 61 percent and 22 percent, respectively. Whitlock's investigation of this drop concluded that, in nearly all cases, Army Reserve lieutenants had, in fact, possessed the required baccalaureate degree. Furthermore, a random survey of the FY 1998 captains on the Army Promotion List (APL) board showed that more than half of these lieutenants not only had their degrees but even had proof of their degrees listed on their Total Army Personnel Data Base–Reserve (TAPDB-R) file, though this is not considered a sufficient source of evidence. Whitlock pointed to an overall lack of communication to the officers that they be prepared to show proof of their degree and a lack of communication between U.S. Army Reserve Personnel Command (AR-PERSCOM, now U.S. Army Human Resource Command [HRC]) offices, that could have helped ensure that the necessary documentation to prove degree status was available.

In more-recent literature, the question of educational qualification remains at issue. To be promoted to captain, officers must have completed both a baccalaureate degree and the Basic Officer Leaders Course (BOLC B).[1] Although the ARNG had increased accessions from OCS in an effort to make up for the shortage of officers coming from ROTC, Feidler (2008) expressed concern about the number of officers commissioned through OCS who do not have their baccalaureate degrees, though he acknowledged that the ARNG is working to resolve this problem. Erlandson (2009) described the limitations of the

---

[1] BOLC is a two-phase course. BOLC A is completed during precommissioning training (e.g., ROTC or OCS). BOLC B is the leadership and branch training section of BOLC, for those who are already officers.

U.S. Army Training and Doctrine Command (TRADOC) in meeting the necessary officer throughput, pointing to the lack of available seats for BOLC II (now part of BOLC B).

At the same time, Reno (2008) was troubled by the decreasing number of graduates of the nonresident Command and General Staff Officers Course, Reserve Component (CGSOC-RC) from the USAR. In his estimation, with a low fill rate of captains and a decreasing number of CGSOC-RC graduates, the ability to promote sufficient numbers to the position of major and higher was a very serious concern for the USAR at the field-grade rank—more serious than the shortfall of captains at the company-grade level.

Feidler (2008) argued that deployments have prevented lieutenants from completing their education requirements. When his report came out, 700 junior officers had recently been passed over for promotion because of a lack of educational progress due to deployment. Feidler further described how deployment could prevent a qualified officer from being promoted if that officer had been cross-leveled and deployed with another unit. While the officer was deployed and ineligible for promotion until returning to the unit, those officers who were not deployed were able to fill the promotion vacancy, leading to a "peer-in-the-rear" promotion that Feidler believed significantly diminishes morale and discourages officers to volunteer for deployments (Feidler, 2008, p. 8).

Erlandson (2009) noted that there are also geographic constraints to promotion in the RCs. There must be a position available in a reserve officer's region in order for him or her to be promoted.

He also argued that limited benefits in comparison to those offered by the other components and the strain of civilian employment hindered the ability of the USAR to effectively retain valuable officers, although he presented no data to support his claim that retention is problematic.

Problems Resulting from the Shortfall

Study authors have contended that the captain shortfall negatively affects individual and unit readiness. At the individual level, there is concern about the need to promote all "fully qualified" officers, as opposed to only the "best qualified."[2] In particular, Reno (2008) argued that an inevitable degradation of officer quality resulting from the less-stringent promotion policy is the most-critical threat to the Army as a whole.

At the unit level, some have argued that ongoing cross-leveling, in response to officer shortages, negatively affects unit cohesion and increases wear on the force. Whitlock (2006) described the strain brought on by the perpetual cross-leveling necessary to man the Army Force Generation (ARFORGEN) model in the RCs. The officer short-fall, Whitlock argued, necessitates a constant cross-leveling of officers and equipment to fill vacancies in deploying units, leading to vacancies in the remaining units. When these remaining units are called up, in turn, the vacancies created by previous cross-leveling must be filled by cross-leveling from other units. This frequent cross-leveling, Whitlock suggested, could result in a lack of effective unit cohesion, affecting the ability of the unit to train and prepare effectively in the reset/train and ready pools of the ARFORGEN model.

---

[2]   Guided by Army Regulation (AR) 600-8-29, Officer Promotions, *fully qualified* means qualified "professionally and morally to perform the duties expected of an officer in the next higher grade" (p. 19). *Best qualified* means fully qualified officers who "meet specific branch, functional area or skill requirements" (p. 19).

Promotion boards will do the following:

(3)(a) The "fully qualified" method when the maximum number of officers to be selected, as established by the Secretary, equals the number of officers above, in, and below the promotion zone. Although the law requires that officers recommended for promotion be "best qualified" for promotion when the number to be recommended equals the number to be considered, an officer who is fully qualified for promotion is also best qualified for promotion. Under this method, a fully qualified officer is one of demonstrated integrity, who has shown that he or she is qualified professionally and morally to perform the duties expected of an officer in the next higher grade. The term "qualified professionally" means meeting the requirements in a specific branch, functional area, or skill.

(3)(b) The "best qualified" method when the board must recommend fewer than the total number of officers to be considered for promotion. However, no officer will be rec-ommended under this method unless a majority of the board determines that he or she is fully qualified for promotion. As specified in the MOI [memorandum of instruction] for the applicable board, officers will be recommended for promotion to meet specific branch, functional area or skill requirements if fully qualified for promotion. (p. 19)

A 2007 Defense Science Board report argued that the RC force structure should be reviewed. The task force concluded that, even with increased authorizations, the current deployment tempo made the mandated five-year dwell time unachievable (Defense Science Board, 2007). Experts interviewed by the task force expressed concern that wear on the force due to the deployment tempo could be hard to repair.

Finally, Whitlock (2002) warned that inattention to the problem would lead to an increased shortfall that would affect higher ranks over time. Indeed, for the USAR, Reno (2008) suggested that the shortage of majors, not captains, was actually the more-problematic issue. Although the USAR could access enough lieutenants over time to fill captain slots, it would take longer to fill slots for majors due to the current captain shortfall, the length of time needed to develop majors, and the difficulty captains have in fulfilling their military educational requirements.

Summary

These studies attribute the initial cause of the company-grade officer shortfall to the reduction of officer accessions in the early 1990s, coinciding with the military drawdown and the challenges presented during initial ROPMA implementation. Further compounding the problem is the fact that ROTC failed to meet mission requirements for nearly a decade and a half, necessitating recruiting from the RC officer pool to adequately man the active force. Additionally, the high operating tempo (OPTEMPO) exacerbated by the personnel shortage and obstacles along the officer career path, such as the management of military education requirements, could further complicate the Army RCs' ability to promote enough qualified officers in a timely manner. Concerns stemming from the lack of captains include the impact on individual performance from years of promoting those who are "fully qualified" rather than "best qualified" and the impact on unit performance from continuous cross-leveling. There are also concerns that this problem is migrating up to the rank of major.

Study Goals and Methods

There are four goals of this study, based on the directives in the House report:

1. Confirm the magnitude of the company-grade officer shortfall in the RCs of the U.S. Army.
2. Identify recommendations that address the company-grade officer shortfall, specifically the shortfall of captains.
3. Assess whether the concept of a National Guard academy is a feasible partial solution to the company-grade officer shortage.
4. Assess the impact of mandating state OCS programs to require candidates to hold a four-year degree to participate in the program.

Although mixed methods were used to achieve these goals, this study is primarily quantitative. To confirm the magnitude of the shortfall, we analyzed the Reserve Component common personnel data system (RCCPDS), Structure and Manpower Allocation System (SAMAS) Master Force (MFORCE), inventory and authorization data provided by the USMC, and a RAND-built unit cohesion file, which is a person-month data set with unit, pay grade, job, demographics, salary, and deployment information for every soldier who served in the military between 1996 and June 2009. It is updated twice per year on five sets of files that RAND receives from the Defense Manpower Data Center (DMDC):

- Defense Enrollment Eligibility Reporting System (DEERS) (demographics)
- work experience (WEX) file (unit, grade, military occupational specialty [MOS]/branch info)
- global war on terrorism (GWOT) file (used for deployment information)
- Reserve pay file: e.g., basic pay, drilling pay, hostile fire
- active pay file: e.g., basic pay, drilling pay, hostile fire.

We also used these data files to conduct inventory projection modeling to calculate the near- and long-term effects of altering input, promotion, and continuation rates.

For some of our analyses, we relied on data we received directly from one of the three Army components. These data include

- AC-to-RC transfer data from the Army's HRC
- ROTC missions from Cadet Command
- numbers of officers directly commissioned in the USAR from Army Reserve Careers Division (ARCD)
- BOLC wait times from Army component manpower or personnel staff officer (G-1)
- USAR captain slots filled by those in other ranks from the U.S. Army Reserve Command (USARC).

We conducted our analyses on the total RC population, including individual mobilization augmentees (IMAs), full-time Active Guard and Reserve members (AGRs), and officers in the special branches (medical, legal, and chaplain). We recognize that, by excluding different subpopulations, we could have arrived at different conclusions. For example, when we exclude the special branches, the observed shortfalls are lower in the USAR (i.e., less problematic) and higher in the ARNG (i.e., more problematic).

To generate ideas for addressing the shortfall, we relied on our internal data analysis and on suggestions captured in 25 interviews that we conducted with more than 50 participants. Participants included representatives from the Department of the Army (DA); Army G-1 and subordinate commands; the National Guard Bureau; the ARNG; five adjutant generals (TAGs) and their staffs; the Office of the Chief of the Army Reserve (OCAR); Army Reserve, G-1; USARC; and ARCD. We also interviewed the USMC Reserve Affairs staff about their company-grade officer shortfall.

Organization of This Monograph

There are four subsequent chapters in this monograph. Chapter Two explores the fill rates for company-grade drilling reservist officers and majors. It also discusses the problems associated with low fill rates and concludes by projecting fill rates into the future. In Chapter Three, we present policy options to improve the AC-to-RC transfer rate, as well as accessions through ROTC, the Officer Direct Commission (ODC) program, and OCS. We address the need for a National Guard academy in this chapter, as well as the question of whether to require state OCS entrants to hold a baccalaureate degree. In Chapter Four, we discuss promotion and retention in the ARNG and USAR. Chapter Five presents conclusions and recommendations on the captain shortfall.

CHAPTER TWO

Company-Grade Officer Fill Rates

This chapter presents data on authorizations and fill rates for company-grade officers in both Army RCs. We also present fill rates for majors, given the concerns expressed in the literature about major inventories. We include data on the USMC Reserve and U.S. Navy Reserve because their RCs are also experiencing a shortfall in company-grade officers. Based on DMDC data, neither the U.S. Air National Guard nor the U.S. Air Force Reserve has a shortfall in company-grade officers.

Company-Grade and Midgrade Officer Fill Rates

In FY 2009, all three Army components met their end-strength objectives. The AC Army met its end-strength objective of 549,015 and requested a temporary increase in this number. The AC was above authorizations for both lieutenants and captains. Both the USAR and the ARNG also met their end-strength objectives (of 205,297 and 358,391, respectively). The ARNG had to reduce its numbers from 368,727 to the congressionally mandated level (per 10 U.S.C. Chapter 1201). The two Army RCs have about the same numbers of officers but different ratios of officers to enlisted members, with the ARNG enlisting about twice as many soldiers as the USAR does.[1]

---

[1] End-strength objectives are based on numbers provided in the FY 2009 U.S. Army Profile, which uses data from the Army and the DMDC, valid as of September 30, 2009. See Headquarters, Department of the Army, 2009, for further detail.

However, in both Army RCs, there has been a persistent shortfall in the number of captains, compared to the authorizations set by the Army. In both RCs, there are more captains authorized than lieutenants in units. According to 2009 SAMAS data, in FY 2009 in the USAR, there were 4,094 lieutenants, 11,332 captains, and 9,754 majors authorized. In the same year in the ARNG, there were 9,254 lieutenants, 12,884 captains, and 8,579 majors authorized. These structures are illustrated in Figures 2.1 and 2.2. In theory, these diamond structures can be maintained if a sufficient number of captains transfer into the RC from the AC and if captains then remain at the captain rank for a longer period of time than officers serve as lieutenants.

Figure 2.1
U.S. Army Reserve Officer Authorizations Illustrated

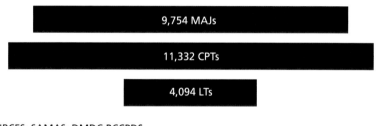

SOURCES: SAMAS; DMDC RCCPDS.
NOTE: MAJ = Army major. CPT = Army captain. LT = Army lieutenant.
RAND MG1045-2.1

Figure 2.2
Army Reserve National Guard Officer Authorizations Illustrated

SOURCES: SAMAS; DMDC RCCPDS.
RAND MG1045-2.2

However, historical and current data demonstrate that the captain fill rate in both the ARNG and the USAR has fallen short of 100 percent for at least the past seven years.[2] Figures 2.3 and 2.4 show how USAR and ARNG authorizations for 2009 (in red) compare to 2009 inventories (in green). The lieutenant fill rate exceeds 100 percent in both the USAR and ARNG. The captain and major fill rates fall below 100 percent, with the red shading representing empty slots.

Figures 2.5–2.8 present the fill rates for lieutenants and captains in both Army RCs from 2003 to 2009. Fill rates are noted above the bars, representing inventory, and the lines, representing authorizations.

Figure 2.3
U.S. Army Reserve Inventory Illustrates Shortfall

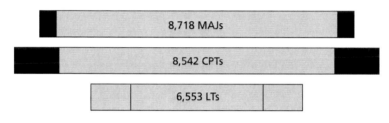

8,718 MAJs

8,542 CPTs

6,553 LTs

SOURCES: SAMAS; DMDC RCCPDS.
RAND MG1045-2.3

Figure 2.4
Army Reserve National Guard Inventory Illustrates Shortfall

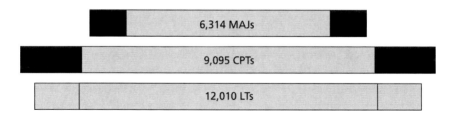

6,314 MAJs

9,095 CPTs

12,010 LTs

SOURCES: SAMAS; DMDC RCCPDS.
RAND MG1045-2.4

---

[2] The literature supports the notion that this shortfall has existed for about 15 years. We examined authorization data from only 2003–2009.

Figure 2.5
Army National Guard Unit Lieutenant Fill Rates and Inventory Versus Authorization Levels, 2003–2009

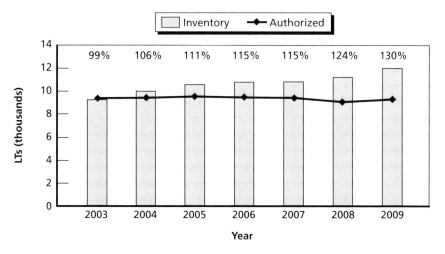

SOURCES: SAMAS; DMDC RCCPDS.
RAND *MG1045-2.5*

Although, as these figures demonstrate, the captain fill rate has not exceeded 75 percent for the past seven years in both RCs, the company-grade officer fill rate is improving. In the ARNG, the lieutenant fill rate has exceeded authorizations since 2004, and the captain fill rate has improved since 2006. The FY 2009 overall ARNG company-grade officer fill rate was 95 percent, with a 71-percent captain fill rate. The captain fill rate has been increasing since 2006, primarily as a result of an increasing inventory. In the USAR, the lieutenant fill rate has exceeded authorizations since at least 2003, and the captain fill rate has improved since 2005; the FY 2009 fill rate was 75 percent. The USAR captain fill-rate increase is due to both an increasing inventory and decreasing authorizations. The overall FY 2009 USAR company-grade officer fill rate was 98 percent.

Because the ARNG commissions many more lieutenants but authorizes a similar number of captains, it is somewhat surprising that the USAR captain fill rate is higher than the ARNG captain fill rate. It could be that USAR has been more successful with directly commis-

Figure 2.6
Army National Guard Captain Fill Rates and Inventory Versus Authorization Levels, 2003–2009

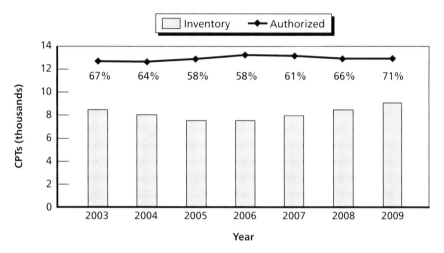

SOURCES: SAMAS; DMDC RCCPDS.
RAND *MG1045-2.6*

sioning captains, as well as bringing them in from the AC, as data we present later will indicate.

There are also company-grade officer shortfalls in the USMC RC and the U.S. Navy RC. In 2009, USMC Reserve company-grade officer authorizations equaled 1,799. According to data provided by the USMC, the total SELRES strength in that year was 887 company-grade officers, resulting in a company-grade officer fill rate of 49 percent. According to DMDC data, the U.S. Navy Reserve company-grade authorizations were 5,165, and inventory equaled 4,698. The company-grade officer fill rate was therefore 91 percent in the U.S. Navy Reserves. In the USMC RC, major fill rates reportedly exceeded 100 percent in 2009, while, in the U.S. Navy RC, according to DMDC data, the major fill rate was 93 percent.

In the Army RCs, the fill rate for majors in both RCs has been declining. Figures 2.9 and 2.10 present the fill rates for majors in the ARNG and USAR. In FY 2009, the major fill rate in the ARNG was 74 percent, and it was 89 percent in the USAR. In the USAR, the

Figure 2.7
U.S. Army Reserve Lieutenant Fill Rates and Inventory Versus Authorization Levels, 2003–2009

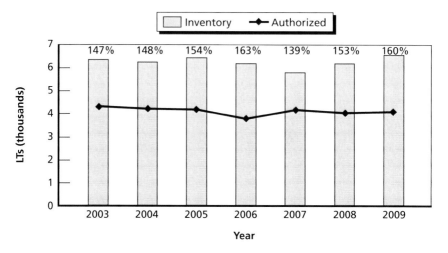

SOURCES: SAMAS; DMDC RCCPDS.
RAND MG1045-2.7

major fill rate has declined by about 30 percentage points over seven years.

Figure 2.11 compares fill rates for lieutenants, captains, and majors in both RCs. At the rank of major, the gap between inventory and authorizations has grown over the past seven years, while the lieutenant and captain fill rates have improved. The lines represent the fill rates.

Fill Rates by Unit Type

Interviewees encouraged us to look at fill rates by unit type, and, indeed, we observe variation here. We examined captain fill rates in both RCs by unit type (represented by standard requirement code [SRC]) from 2003 to 2009. Specifically, we looked at the authorization and fill rates of individual captains within units. The total number of units in each "unit type" varied. For example, in 2009 in the USAR, there were

Figure 2.8
U.S. Army Reserve Captain Fill Rates and Inventory Versus Authorization Levels, 2003–2009

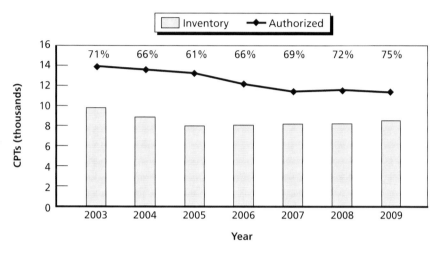

SOURCES: SAMAS; DMDC RCCPDS.

RAND *MG1045-2.8*

917 individual authorizations for 37 authorized civil affairs units and 459 individual authorizations for 99 authorized engineering units.

We noticed several patterns across this 2003–2009 time frame. First, the captain fill rate varied a great deal from unit to unit. For example, in 2009 in the USAR, there was a 40-percent captain fill rate in civil affairs units and an 80-percent captain fill rate in medical units. Second, fill rates vary across time. For example, in civil affairs units in the USAR in 2003, there was a 70-percent captain fill rate. By 2009, as noted above, this fill rate had dropped to 40 percent. Authorizations increased during this time period from 626 to 917, which could help to explain the drop in the fill rate. However, USAR chemical units in 2003 had an 80-percent captain fill rate. By 2009, authorizations for captains in these units had dropped from 154 to 104, but the fill rate had also gone down to 60 percent. Changes in authorization requirements in the future force structure are likely to continue to affect unit fill rates in the out-years.

Figure 2.9
Army National Guard Major Fill Rates and Inventory Versus Authorization
Levels, 2003–2009

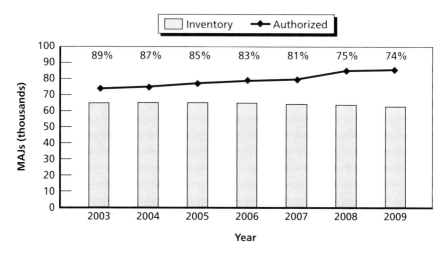

SOURCES: SAMAS; DMDC RCCPDS.
RAND MG1045-2.9

   Table 2.1 shows the types of units, separately for the ARNG and USAR, whose FY 2009 fill rates were less than or equal to 50 percent, between 50 and 70 percent, and equal to or greater than 70 percent. More units in the USAR have a less than 70-percent captain fill rate than had a rate equal to or higher than 70 percent. The appendix provides the full names of all the unit types listed.

Problems Resulting from Shortfalls

The unit fill-rate data, as well as the literature, suggest that unit readiness could be compromised by a shortage of captains. We attempted to verify this hypothesis through primary and secondary data analysis. We explored whether the following problems exist in the ARNG and USAR as a result of an insufficient number of captains:

- Units are less likely to be deemed deployable.
- There is more cross-leveling due to the shortfall.

Figure 2.10
U.S. Army Reserve Major Fill Rates and Inventory Versus Authorization
Levels, 2003–2009

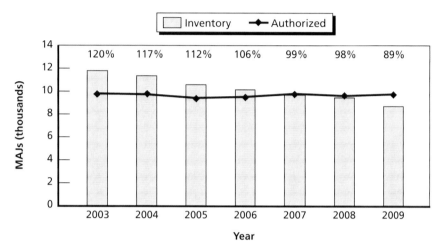

SOURCES: SAMAS; DMDC RCCPDS.
RAND *MG1045-2.10*

- Promotion to major is slowed.
- Lieutenants have insufficient time to develop before they lead units.
- Captains are deployed more frequently than are other officers.

Although the literature also speculates that, through "fully qualified" promotion practices, the quality of company-grade officers could be worsening, we did not attempt to verify this hypothesis.

Interviewees disagreed that units are less likely to be deemed deployable, but they did acknowledge that unit readiness is achieved by cross-leveling. Interview data were inconclusive on the extent to which cross-leveling occurs because of the shortfall in captains. Certainly, cross-leveling happens in response to personnel readiness needs, but those needs include not only captains but also having the appropriate MOS-qualified senior staff.

Although one interviewed TAG did admit to keeping officers in the captain grade for the maximum time in grade (TIG), our data analysis indicates that time to major has not increased over time in the

Figure 2.11
Changes in Fill Rates from 2003 to 2009 at the Lieutenant, Captain, and Major Ranks in the Army National Guard and U.S. Army Reserve

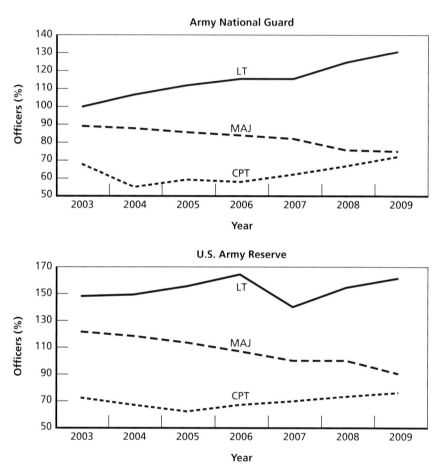

SOURCES: SAMAS; DMDC RCCPDS.
RAND MG1045-2.11

ARNG. Because the maximum TIG from captain to major is seven years in both RCs, we looked at the number of months until promotion to major for new captain cohorts in 1997–2002. For ARNG captains, months to major declined from 71 months for the 1997 new captain cohort to 64 months for the 2002 new captain cohort. For USAR

Table 2.1
Fiscal Year 2009 Army National Guard and U.S. Army Reserve Units at Various Captain Fill Rates

Fill Rate	ARNG	USAR
≤50%	HQ, MD, ME, MI, SB	AV, CA, CS, ME, TC
50–70%	CM, CS, IO, PI, SF	CM, EN, FI, IO, MI, PI, PO
≥70%	AD, AG, AQ, AR, AV, EN, FA, FI, IN, MP, OD, QM, SC	AG, CH, HQ, IN, JA, MD, MP, OD, QM, SC, SP, TDA

SOURCES: SAMAS; DMDC RCCPDS.
NOTE: Common unit types are adjutant general (AG), aviation (AV), chemical (CM), composite service (CS), engineer (EN), financial management (FI), headquarters (HQ), infantry (IN), information operations (IO), medical corps (MD), mechanized (ME), military intelligence (MI), military police (MP), ordnance (OR), public affairs (PI), quartermaster (QM), and signal corps (SC). ARNG unit types are air defense artillery (AD), contingency contract support battalion (AQ), armor (AR), field artillery (FA), reconnaissance and support battalion (SB), and special forces (SF). USAR unit types are civil affairs (CA), chaplain (CH), judge advocate (JA), psychological operations (PO), medical specialist corps (SP), transportation (TC), and table of distribution and allowances (TDA).

captains, months to major did increase from 50 to 56 months, but this increased average TIG falls far below the maximum of 84 months.

Interviewees disagreed that lieutenants are put into leadership positions with insufficient training and experience. As Figure 2.12 demonstrates, 36 percent of captain slots are filled by lieutenants (9 percent second lieutenants and 27 percent first lieutenants). Fewer (9 percent) are filled by majors. When we asked ARNG and USAR leaders about their experiences in assigning lieutenants to captain slots, they replied that they have sufficient numbers to assure them that they can appoint strong leaders.

Finally, we did not observe higher deployment rates for captains than for lieutenants or majors. We expected to see higher rates of deployment for individual captains. We used the RAND unit cohesion file data to calculate the percentage of lieutenants, captains, and majors deployed in each month, October 15, 2000, through December 15, 2009. For the ARNG, in 76 of these 111 months, first lieutenants had the highest deployment rate. In only 38 months did captains have the highest deployment rate. Looking at this same population in the

Figure 2.12
**Proportion of U.S. Army Reserve Captain Slots Filled by First and Second
Lieutenants in January 2010**

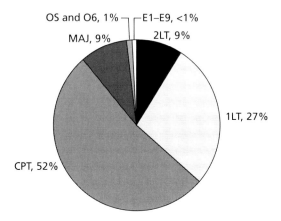

SOURCE: USARC, Accessions branch: ATTRS.
NOTE: O5 = pay grade for Army lieutenant colonel. O6 = pay grade for Army colonel.
E1 and E2 = pay grades for Army privates. E3 = pay grade for Army private first class.
E4 = pay grade for Army corporal or specialist. E5 = pay grade for Army sergeant.
E6 = pay grade for Army staff sergeant. E7 = pay grade for Army sergeant first class.
E8 = pay grade for Army master sergeant or first sergeant. E9 = pay grade for Army
sergeant major or command sergeant major.
RAND MG1045-2.12

USAR over the same 111 months, in 70 of these months, majors had
the highest deployment rate. In only 11 of these months, captains had
the highest deployment rate.

Given the magnitude of the shortfall in captains, we were sur-
prised by our interviewees' inability to readily describe resulting problems. When we asked USMC interviewees to describe the problems
resulting from their low captain fill rates in their RC, they responded,
"For so long, we have just been asked to press on as is that we don't
really know the answer to the question of how bad it really is." It could
indeed be the case that this problem has been going on for so long now
that alternative solutions (e.g., filling captain slots with lieutenants)
have become the norm and Army RC leaders have stopped considering
the shortfall a pressing problem.

Inventory Projection Modeling

As mentioned above, we conducted modeling to project future inventories of officers. Our goal for the modeling was to ascertain how long it would take to get to a 100-percent captain fill rate in each Army RC. We built synthetic cohorts of officers from the second lieutenant to the lieutenant colonel rank in both RC. We based the number and composition of these cohorts on DMDC RCCPDS inventory data and SAMAS authorization data from 2001–2009. In our projections, we used 2016 authorization numbers (which is as far out as we have authorizations data) for 2017 and beyond.

We built ten models to test various accession and promotion assumptions. In each model, we held continuation rates to the average we observed over the past ten years. Here, we describe our continuation, accession, and promotion assumptions.

Continuation Rates

As mentioned above, we assumed that the average continuation rates we observed in the data over the past ten years would be applicable for future officer cohorts. The average year-to-year loss rate we observed over a ten-year period is 8.61 percent for the ARNG and 10.39 percent for the USAR.[3] We estimate different separation behavior for each officer in the model. The loss for one year is calculated as the number of separations in that year (net of promotions) as a percentage of the number of officers in the grade the previous year (to avoid counting new accessions in the current year). As the model runs, each officer has at first a small probability of separation that changes over time based on time in service (TIS) (increasing the probability) and potential for promotion (decreasing the probability). The loss rate provides the numerical probability of separation for a period of service for each grade; every grade has a distinct loss rate, so an officer with four years

---

[3]   Loss rates have been declining from peaks in 2004–2005 (for ARNG) and 2005 (for USAR) on the order of 2 percentage points for ARNG and 4 percentage points for USAR each year. Average yearly loss rates observed in the data are 8 percent at O1 and O2 (pay grades for Army second and first lieutenants, respectively); 11 percent at O3 (pay grade for Army captain); 9 percent at O4 (pay grade for Army major), and 15 percent at O5.

in grade as first lieutenant would not face the same chance of separation as four years in grade as a captain, and so on.

Accessions

Accessions are the number of new officers introduced into the model in each year at all ranks between second lieutenant and lieutenant colonel. These new officers could have been commissioned as new second lieutenants, or they could have transferred in from the AC, the IRR, or even another service. They could also have been directly appointed to any of the ranks in our model. We base the proportion of officers entering the model at each rank on historical observations.

We observed a 2-percent and a 4-percent increase in accessions in the ARNG and USAR, respectively, over the past ten years. However, accessions have increased over the past four years. For example, in 2009, the ARNG commissioned 57 percent more from the ROTC scholarship program, 7 percent more from the ROTC nonscholarship program, 32 percent more from the state OCS programs, and 68 percent more through direct commissioning than they had in 2006.

Therefore, we increased the accession rates in the projection modeling using two scenarios. In the first scenario, we assume an annual, exponential 4-percent increase in accessions in the ARNG and an annual, exponential 6-percent increase in accessions in the USAR. We based these numbers on the highest yearly percentage change in accessions demonstrated over the past ten years for any source of commission (normalized by the underlying population), which is lower than the total percentage change in accessions driven by all sources of commission. The gain rate by source of commission varies. In order to reflect gains by individual source of commission in proportion to historical trends, it is necessary to calculate the contribution to overall accession growth that each source represents. For example, in 2007, the ARNG gained about 100 more new officers than they had in 2006 from both the ROTC scholarship and the state OCS programs. However, due to variation in the overall numbers, there was an increase of 32 percent from ROTC and an increase of 11 percent from the state OCS commissioning source. To arrive at an average increase of 4 percent, we normalized the 2006 to 2007 data to account for the number of officers gained by source of commission. For instance, if there were two sources and one source had an increase of 20 percent and accounted for 10 percent of the officer gains from all sources, we assigned a 2-percent increase to this source. If the other source had a gain of 10 percent and accounted for the remaining 90 percent of the gains, we assigned it a 9-percent increase. The sum of these increases, i.e., 11 percent, would be the percentage increase from all sources. Note that the highest gain from any source—here, the 9-percent increase—is less than the total percentage gain.

In the second scenario, we assume a 10-percent increase in accessions in both RCs in 2010, relative to the 2009 accessions. In 2011, we assume a 20-percent increase in accessions, relative to the 2009 numbers, and in 2012, a 30-percent increase in accessions, relative again to the 2009 numbers. For 2012 going forward, we assume no increases in accessions. Achieving these early high growth rates assumes that the economy will remain favorable to recruiting efforts and that the Army can temporarily direct greater resources to accessions to increase the captain fill rate in its RCs.

Promotions

We also modeled two scenarios on time to promotion to captain. Similar to the process for separation, each officer is considered individually for promotion, developing a within- and across-grade distribution of officers selected for promotion defined by parameters that represent service-wide policy (most notably, an officer's TIG and overall promotion opportunity by grade). Across both RCs, an officer can now be promoted from a second lieutenant to a captain in three years and five months. However, recent data demonstrate that most lieutenants are in second and then first lieutenant grade for at least five years. We tested the effects of accelerating this time frame. In one model, we assume an average of four years TIG as a second and first lieutenant. However, both RCs have the capacity to promote officers to captain earlier if they use a promotion vacancy board (PVB). Indeed, the ARNG has relied on the PVB system for some time now, while the USAR has rarely exercised its authority to do so. The USAR has been encouraged by the Army to increase its use of these vacancy promotion boards. Therefore, in another model, we varied TIG for promotion. We assumed that, in 2011 and going forward, the USAR would promote 10 percent of second lieutenants to first lieutenant and 10 percent of first lieutenants to captain in a PVB system, which requires less TIG. In this same model, we assume that the FY 2009 percentages of ARNG lieutenants promoted to captain through the unit vacancy boards (UVBS) will continue.

We also modeled two scenarios on promotion to major. We observed that, over the past ten years, it took, on average, five years to be promoted from captain to major in both RCs. We therefore used this number going forward. We also held captains in place for seven years in a separate model, to test the extent to which such a change in practice would accelerate achieving a 100-percent captain fill rate.[4]

In total, we ran ten different models. Modeling results are displayed in Tables 2.2, 2.3, and 2.4. Depending on the model, the captain fill rate would reach 100 percent in the next five to nine years in the USAR and 6.5 to 10.5 years in the ARNG. The tables also demonstrate that the year at which the captain fill rate will reach 90 and

Table 2.2
Projected Fill Rates for Captains, Based on Rapid Initial Increase Followed by Leveling Off

	Fill-Rate Timing					
	90%		95%		100%	
Model	O3	O4 Fill Rate	O3	O4 Fill Rate	O3	O4 Fill Rate
1: ARNG	2 years	78%	3.5 years	89%	6.5 years	96%
2: USAR	1 year	80%	3 years	92%	5 years	98%

NOTE: Leveling off is defined as year t + 1 accessions = year t × 1.1, year t + 2 accessions = year t × 1.2, year t + 3 accessions = year t × 1.3, years t + 3…n accessions = year t × 1.3.

---

[4] In all models, we assume a 65-percent promotion rate from lieutenant to captain and from captain to major, a 60-percent promotion rate from major to lieutenant colonel, and a 35-percent promotion rate from lieutenant colonel to colonel.

Table 2.3
Army National Guard: Projected Fill Rates for Captains, Based on 4-Percent Exponential Growth Yearly and Varied Promotion Rates

| | Fill-Rate Timing | | | | | |
| | 90% | | 95% | | 100% | |
Scenario	O3	O4 Fill Rate	O3	O4 Fill Rate	O3	O4 Fill Rate
Status quo: 4 yrs to CPT, 5 yrs (min) to MAJ	4.25 yrs	90%	7.5 yrs	94%	10.5 yrs	95%
Model 4: 3 yrs to CPT, 7 yrs (min) to MAJ	3.5 yrs	88%	6 yrs	92%	8.75 yrs	94%
Model 5: 3 yrs to CPT, 5 yrs (min) to MAJ	4 yrs	87%	7.75 yrs	92%	9 yrs	94%
Model 6: 4 yrs to CPT, 7 yrs (min) to MAJ	4 yrs	91%	6.5 yrs	92%	9 yrs	92%

95 percent for the ARNG and USAR in each of the different models. It also shows the major fill rate in those years.

Our modeling demonstrates that, under a rapid-increase scenario in which accessions increase by 10–30 percent over the next three years followed by a leveling-off period, the captain fill rate will reach 100 percent in 6.5 years in the ARNG and five years in the USAR. The major fill rate would be 96 and 98 percent, respectively, under these assumptions. This modeling also assumes historical average continuation and promotion rates. Furthermore, under these scenarios, the captain fill rate could increase to 90 percent almost immediately—in one year in the USAR and in two years in the ARNG. Recruiting 10 percent more officers in 2010, followed by 20 percent more than the 2009 numbers the following year and 30 percent more the next, would entail aggressive accessioning policies and practices.

Accessing a net of 4 percent (in the ARNG) and 6 percent (in the USAR) more officers each year is also an aggressive target, but one that

Table 2.4
U.S. Army Reserve: Projected Fill Rates for Captains, Based on 6-Percent Exponential Growth Yearly and Varied Promotion Rates

Scenario	Fill-Rate Timing					
	90%		95%		100%	
	O3	O4 Fill Rate	O3	O4 Fill Rate	O3	O4 Fill Rate
Status quo: 4 yrs to CPT, 5 yrs (min) to MAJ	2.75 yrs	77%	6.25 yrs	83%	9 yrs	92%
Model 8: 3 yrs to CPT, 7 yrs (min) to MAJ	2 yrs	72%	5 yrs	77%	7 yrs	87%
Model 9: 3 yrs to CPT, 5 yrs (min) to MAJ	2.75 yrs	75%	6 yrs	87%	8.75 yrs	94%
Model 10: 4 yrs to CPT, 7 yrs (min) to MAJ	3 yrs	71%	5.5 yrs	76%	7.5 yrs	86%

has been exceeded in recent years. Under these scenarios, the ARNG would achieve a 100-percent captain fill rate in 10.5 years if past average continuation and promotion rates were maintained. The USAR would achieve a 100-percent captain fill rate in nine years if historical average continuation and promotion rates were similarly maintained. When the captain fill rate reaches 100 percent, the major fill rate would be 95 percent in the ARNG and 92 percent in the USAR.

The promotion rates used in our other models would require different approaches than are now used in the RCs. Some of our modeling assumes that captains stay in grade for an average of seven years (functionally, this means opening in-zone promotion windows only past the sixth year of service), which is, on average, two years longer than the 2002 new captain cohort remained in captain grade. By retaining captains in grade for these two extra years, the captain fill rates would increase to 100 percent about 1.5 years earlier in both RCs than it would under models assuming historical promotion to captain rates.

However, the major fill rate would not rise as quickly as it does when promoting captains to majors using historical rates. The major fill rate would be 92 percent in the ARNG and 86 percent in the USAR, compared to the 95- and 92-percent rates observed in models based on historical promotion rates.

Similarly, promoting lieutenants to captains using more-aggressive vacancy promotion boards would entail both RCs achieving a 100-percent captain fill rate about six months earlier than they do without using these boards. In these models, the major fill rate is 94 percent for both RCs by the time the captain fill rate reaches 100 percent. This major fill rate is lower by 1 percentage point in the ARNG and higher by 2 percentage points in the USAR than we observe in our models using historical promotion rates.

The most-aggressive model results in a 100-percent captain fill rate in the ARNG in 6.5 years and in five years in the USAR. In this model, we assume that accessions increase by 10–30 percent over the next three years, followed by a leveling-off period. This modeling assumes historical average continuation and promotion rates. (If lieutenants were promoted to captain more quickly, the captain fill rate could reach 100 percent even earlier.) Recruiting 10 percent more officers in 2010, followed by 20 percent more than the 2009 numbers the following year and 30 percent more the next, would entail aggressive accessioning policies and practices, but it could be worth investing in such policies and practices in the short term, particularly given the current unemployment context, to dramatically boost the captain fill rate.

A more-conservative model results in a 100-percent captain fill rate in the ARNG in 10.5 years and in nine years in the USAR. In this model, accessions increase by 4 percent in the ARNG and 6 percent in the USAR each year for the next 10.5 years. Recent average loss rates continue. Lieutenants are promoted to captain after serving as a second and then a first lieutenant for a total of four years. On average, captains are promoted to major after five years, which is the current average TIG.

Even using fairly aggressive assumptions, the captain fill rate is not likely to approximate 100 percent in the next four years. There could be, as one interviewee noted, "no way to fix the problem, unless you look at structure reorganization as part of the overall solution." Indeed, TRADOC, at the direction of the Chief of the U.S. Army, is now analyzing the extent to which current major and lieutenant colonel billets could be replaced with noncommissioned, warrant, or lower-ranked officer billets. This exercise could affect structures for lower ranks and provide an opportunity to review captain authorizations. It could be useful to ascertain what kinds of captain positions are being filled by first lieutenants, for example.

However, even if the structure is changed, it is likely that the requirements for captains will continue to exceed inventory. Chapter Three discusses policy options to improve affiliations and accessions into the Army RCs. Chapter Four presents options for improving promotion and retention rates.

Affiliations and Accessions

This chapter presents recommendations to increase affiliations and accessions of officers into the Army RCs. The current economic climate should facilitate efforts to improve affiliations and accessions, but some additional policy changes could be beneficial. We first discuss the AC-to-RC transition. All U.S. Army officers sign an initial eight-year service contract upon entry into the military. Typically, the contract specifies that some initial period of service will be in the AC (two, three, or four years) with the remaining obligation served in an RC. Indeed, under the current authorization structure, some transfers of company-grade officers from the AC to the RC are needed to ensure that the RCs will have sufficient numbers of captains and higher-ranking officers.

The RCs also grow their own officers. Reno (2008) and our interviewees agreed that, although there are currently more lieutenants than authorized, it is important for the RCs to continue increasing inventories of lieutenants so that there will be sufficient numbers to promote to captain. Although there are concerns that continuing to recruit lieutenants could push the ARNG in particular over its end-strength limit, the ARNG is planning to continue to aggressively accession new lieutenants in the next few years. (It is also proposing an exemption from counting officers in training against end-strength caps.)

In this chapter, we explore the three main commissioning sources for the RCs: ROTC, state OCS, and the ODC program (which is relevant for higher officer grades, as well as for lieutenants). In exploring the state OCS source of commission, we address the question from H.R. 111-166 on whether entrants into these programs should hold baccalaureate degrees. We also consider here how establishing a new National Guard academy might affect the company-grade officer fill rates.

Active Component–to–Reserve Component Transitions

According to 10 U.S. Code 651, AC officers who have a remaining military service obligation (MSO) and are otherwise qualified, shall, upon release from active duty, be transferred to an RC of the armed force to complete the service required. Many of these officers end up in the IRR, which does not require regular drilling. Although these officers can be deployed, they do not count as being part of the SELRES.

There are incentives to attract officers coming off active duty into the SELRES, including up to 24 months reduced MSO and a $10,000 affiliation bonus. Officers in an RC are also eligible for tuition assistance. Although the reduced MSO is reportedly a strong incentive, there are no empirical studies testing its effectiveness. Interviewees did not believe that the $10,000 was an effective incentive, but, again, we found no studies testing its effectiveness.

In 2007, the Army Human Resource Command (HRC) was given responsibility for recruiting officers leaving the AC into drilling reserve units in the SELRES. Table 3.1 presents the number of officers who left the AC and affiliated with an RC unit in 2008 and 2009. Reserve affiliations are based on data from HRC and are exclusive of those who joined an RC in 2008 and 2009 but had left the AC in previous years. The table also presents Army G-1 data on the total number of officers who left the AC with less than ten years of service and departure codes that represent eligibility for an RC commission. For example, officers were excluded if they had loss transaction codes representing death, disability, or conscientious objection.

These gains exceeded HRC's goals, and they represent about 40 percent of the potentially eligible pool of officers leaving the AC. In FY 2009, the goal was to recruit 287 AC officers into the ARNG and

Table 3.1
Officers Transitioning from Active Component Army to Reserve Component Units

Year	Leaving AC	Entering ARNG	Entering USAR
2008	2,314	499	561
2009	2,572	462	564

SOURCE: HRC telephone and email communications with the authors.

350 AC officers into the USAR.[1] This goal was reportedly based on FY 2008 goals rather than on targeting a proportion of those leaving the AC. The recruiting goal for FY 2010 was to transition 400 AC officers into the ARNG and 500 into the USAR. As of March 2010, HRC had achieved 75 percent of this target. However, interviewees expressed concern about reaching these goals by the end of the fiscal year.

HRC interviewees described factors hindering their ability to recruit from the AC Army. First, they lack visibility into AC officers' intentions to continue serving in the AC. They are, in their words, "missioned against something we can't see." They do not have sufficient information on whom to contact when. Rather than simply cast a wide net, RC recruiters would prefer to contact individual AC officers who are thinking of leaving the AC. Moreover, they are sometimes unable to detect vacancies. For example, if a lieutenant is filling a captain slot, that position might not appear to be vacant, although a captain transferring from the AC could indeed fill this slot. Compounding this lack of visibility is the reported tendency for AC officers not to speak with counselors until they are at the point of out-processing. All AC officers are required to speak with an RC counselor before they are out-processed, but the timing of these meetings varies across sites. At the out-processing point, most have made up their minds about next steps, and many lack an understanding of RC options. Interviewees speculated that most AC officers talk primarily with their commanders about their resignation plans. Interviewees asserted that AC command-

---

[1] Although these goals are split by RC, HRC is only provided with target numbers for the total number of AC soldiers it is expected to transition into an Army RC.

ers are unlikely to have full information about RC options for departing officers.

A second barrier involves the process of commissioning an AC officer into an RC. Because an officer cannot have two appointments at once, he or she must resign from the AC before signing an oath for an RC commission. In most cases, AC officers are unable to sign an RC commission before they leave AC service. This requirement challenges HRC to keep track of these officers after they leave AC service and ensure that they sign the oath for an RC commission after their AC commission resignation has been approved.

Several options could increase the number of AC officers transferring to an RC. They include gathering better data on AC officers nearing separation; ensuring visibility into officer vacancies; offering potential AC transfers more counseling on career decisions earlier in the process (an approach also recommended by Feidler, 2008); and offering new incentives to join an RC. In terms of counseling, officers leaving the AC might need additional guidance on how to blend a civilian career with an RC commission. Offering them job opportunities alongside an RC commission could serve as a strong incentive to join an RC. Financial incentives are another option and would have to be competitive with whatever incentives the active-duty officer faces for staying in the AC. Interviewees suggested that graduate school reimbursement would be an additional incentive worth exploring. The Navy Reserve offers financial assistance for advanced degrees in law and medicine, and the USAR already offers "kickers" to supplement GI Bill benefits, as well as financial assistance to repay education loans. A model developed at RAND could simulate the effect of additional programs, such as Graduate School for Active Duty Service Obligation (GrADSO) on RC officer retention (Mattock et al., undated). Others suggested that a sabbatical program for officers leaving the AC to allow them time off from TIG for promotion purposes to finish military or civilian educational obligations while maintaining the ability to drill and gain points toward retirement would be a strong incentive. Finally, interviewees suggested options to allow AC officers to sign an RC oath before they leave AC service. One is to change the statute to allow for a single commission. Another is to allow an exiting AC officer to sign the

RC commission at time of exit but wait to date the commission until the AC resignation is accepted. A fourth option would be to encourage an exiting AC officer to sign a letter of intent to join an RC and award him or her with an increased signing bonus (perhaps targeted to areas of concentration with RC shortages) once the AC resignation is accepted, if the officer follows through on the letter of intent.

Commissioning Sources

We now turn to sources of commissioning new officers into the Army RCs. We focus primarily on ROTC, state OCS programs, and the ODC program. The first two sources of commissioning bring new second lieutenants into the RCs. Although the plurality of officers brought in through direct commissioning are also second lieutenants, this source of commission is used to bring in officers at higher ranks as well.

Reserve Officer Training Corps

ROTC is a major source of new commissions for both the ARNG and the USAR. The U.S. Army Accessions Command has recently been given individual ROTC missions for each of the three components: the active Army, the ARNG, and the USAR. Table 3.2 presents Army G-1 data on the missions for the two RCs, which grow through 2012. Cadet Command reported that, to achieve current target numbers, 45 percent of ROTC cadets should be joining an RC. Cadet Command interviewees believe that this growth can be accommodated; scholarship incentives are reportedly sufficient to attract new cadets needed to meet these growth targets, and the current ROTC infrastructure is sufficient to accommodate the growth.

Interviewees did express some concern, however, that there would be cadets who prefer either the AC or the ARNG to the USAR (because of the lack of combat units in the USAR) and that there would likely be some forced distribution into the USAR.

ARNG interviewees complained about this forced distribution, as well as the order-of-merit list (OML) process, which allows the AC

Table 3.2
Reserve Officer Training Corps Missions for Fiscal Year 2010–Fiscal Year 2012

	FY 2010	FY 2011	FY 2012
ARNG	1,300	1,400	1,500
USAR	745	745	795

SOURCE: Cadet Command communications with the authors.

to commission cadets scoring at the top of this list. Cadet Command reported that there are no plans to change the OML. Interviewees stated that, if there were sufficient incentives for cadets to join an RC, the ARNG and USAR could meet their missions by committing cadets to a Guaranteed Reserve Forces Duty (GRFD) or Dedicated Army National Guard (DEDNG) scholarship track early in their educational career, thus allowing the two RCs to meet their missions before the OML process. There are no statutory caps on the numbers of these scholarships that Cadet Command could offer, though the number will, in practice, be limited by the budget available for such scholarships. However, all three components must meet their individual missions before any one component can recruit beyond its mission.

Interviewees suggested incentives that might attract cadets to an RC early in their educational career. Many stressed that cadets need more and earlier information about the RCs (Feidler, 2008, and Reno, 2008, also expressed this concern). Some of the TAGs we interviewed would like more control over allocating the GRFD and DEDNG scholarships so that they can recruit their own cadets directly. Other interviewees suggested that, at the point of contracting, cadets could be offered monetary bonuses, guaranteed funding for graduate school, choice of branch (not all options could be guaranteed, however, given that not all branches are represented in each region), or guaranteed internships. In general, suggestions focused on incentives that would support cadets in aligning their academic major with future work opportunities. A related suggestion is to give veterans preference rights for federal government civil service positions to those on DEDNG scholarships so that they can compete for civil service positions before they have fulfilled the minimum active duty required to obtain veteran status.

Finally, interviewees suggested that DEDNG scholarship students could be simultaneously eligible for the Montgomery GI Bill–Selected Reserve (MGIB-SR) (10 U.S.C. Chapter 1606). The DEDNG scholarship is available for up to three years but does not allow simultaneous use with the Guard GI bill.[2] Others suggested that students with prior enlisted service using the GI bill should be courted by TRADOC for nonscholarship ROTC slots.

Several interviewees recommended that AC officers who have completed their active-duty service obligation (ADSO) fulfill their remaining MSO in the SELRES rather than in the IRR—an approach also endorsed by Feidler (2008). When questioned about this potential policy change, Cadet Command interviewees speculated that this change would not adversely affect ROTC recruiting. They reported that many cadets are choosing additional active-duty service obligations for graduate school (GrADSO) and for branch-of-choice active-duty service obligations (BrADSOs).

Other, related, options include structuring ROTC cadet contracts so that AC officers who leave active-duty service at the end of their ADSOs are automatically transferred into the SELRES, forcing them to proactively "opt out" if they want to join the IRR instead (and perhaps automatically doing so if they move to a region where there is no need for their occupational specialty). Another option would be to vary service obligations based on incentives awarded during precommissioning. As noted above, ROTC cadets can already lengthen their ADSOs in exchange for choice of branch. Similar options could allow AC cadets, at the time of contracting, to receive incentives in exchange for four years of active duty, two years in the SELRES, and then two years in the IRR, or some other combination that results in less time spent in the IRR.

Several interviewees expressed concern about the Early Commissioning Program (ECP), which is part of ROTC. Under this program,

---

[2] Details about eligibility for the MGIB-SR can be found at Reserve Officer Training Corps, undated.

about 140 cadets graduate from one of five two-year military junior colleges (MJCs) each year and are commissioned as second lieutenants. These officers join reserve units (about 75 percent join an ARNG unit and the rest join USAR units) but are not deployable until they complete their four-year degrees. Although this is a small number of lieutenants, interviewees questioned whether this program is beneficial. Many reported that these early-commissioned lieutenants are not tracked well and that half never complete their baccalaureate degrees, which means that they are not promotable to captain.

Interviewees suggested that, instead of commissioning graduates from the MJCs, Cadet Command should develop two-plus-two programs linking the MJCs with four-year colleges already participating in ROTC. A graduate of one of these programs would then be commissioned when he or she completed his or her baccalaureate degree. If this policy change was implemented, incentives would need to be developed to continue to attract students to the MJCs without the early-commissioning option.

The Officer Direct Commission Program

As is demonstrated in Table 3.3, direct commissioning is a substantial source of new officers for the USAR. The ARNG also uses ODC and directly commissions about one-third of the number of officers that the USAR does each year.

As Table 3.3 displays, there has been an increase in the proportion of second lieutenants coming in through ODC and a decrease in the proportion of captains between 2001 and 2009. In 2001, about 400 captains (260 excluding chaplains and medical branches) came into the USAR via ODC. In 2009, the USAR brought in 160 captains (90 without the professional branches) via ODC.

Both the ARNG and the USAR would like to accession more captains via ODC. Erlandson (2009) and Feidler (2008) endorse this approach. The USAR recently changed the way it rewards a recruiter by crediting him or her for recruiting an officer brought in via ODC in the month in which the officer was recruited, instead of having to wait until the ODC package is approved. In addition, the Army Reserve G-1 has proposed to the DA the option to directly commission to cap-

Table 3.3
The Officer Direct Commission Program in the U.S. Army Reserve, 2001–2009

Total No. Direct Officer Commissions in USAR	2001	2002	2003	2004	2005	2006	2007	2008	2009
	1,267	1,263	1,078	766	535	609	1,082	1,028	1,007
Percentage at each rank									
2LT	26	26	35	42	44	45	49	48	46
1LT	25	22	25	23	18	19	23	24	25
CPT	30	30	25	18	17	17	14	16	16
MAJ	12	17	10	9	12	12	7	8	8
LTC	7	5	5	7	9	7	7	4	4

SOURCE: ARCD communications with the authors.

NOTE: 2LT = Army second lieutenant. 1LT = Army first lieutenant. LTC = Army lieutenant colonel. The numbers include chaplains and medical service branches. Without these professionals, the USAR directly commissioned 831 officers in FY 2009.

tain individuals with certain skill sets. For example, a logistics manager at a private company could be accessioned as a captain into the Army Logistics Corps. These potential officers could have prior service, but it would not be required. Therefore, the proposal limits those without prior service from leading troops for several years. Any ruling on this proposal would cover all three components. Indeed, the ARNG is also considering directly commissioning more civilians as captains in engineering, civil affairs, and military intelligence. This proposal is not supported by all of our interviewees—some disapprove of commissioning a civilian without military training to the rank of captain.

State Officer Candidate School Programs

The state OCS programs are a major source of new second lieutenants for the ARNG. Table 3.4 presents the number of ARNG second lieutenants commissioned via the state OCS programs from 2001 through 2009. Although the number of commissions peaked in 2003, they increased between 2006 and 2009.

Table 3.4
Army National Guard Officers Commissioned Through State Officer Candidate School Programs

	2001	2002	2003	2004	2005	2006	2007	2008	2009
ARNG	776	1,109	1,312	1,092	871	674	752	814	893

SOURCE: DMDC RCCPDS.

H.R. 111-166 requested the Secretary of Defense to examine whether soldiers entering state OCS programs should be required to possess a baccalaureate degree. Current policy dictates that OCS entrants possess 60 or more college credit hours. The motivation for the H.R. 111-166 question stems from the fact that many officers who graduate from a state OCS program never complete their baccalaureate degree and, therefore, cannot be promoted to captain.[3] Table 3.5 presents the proportion of officers commissioned into the ARNG by educational attainment from 2001 to 2009. The proportion commissioned without a baccalaureate degree has declined from 42 percent in 2001 to 25 percent in 2009.

Many of the officers who were commissioned without a degree in 2001 were eventually promoted to captain, which means that they obtained their degrees. Of the 1,425 reservists commissioned as second lieutenants into the ARNG in 2001, 598 came through the state OCS programs without a baccalaureate degree (and one had an "unknown" educational attainment). As presented in Table 3.6, by 2008, 345 had obtained the degree and been promoted to captain (58 percent). In comparison, 72 percent of those who started with a degree were promoted to captain by 2008. Those who started with a four-year degree were statistically significantly more likely to be promoted to captain.

However, if obtaining more captains is critical, it should be worthwhile to incur the cost of training 253 soldiers who leave without making captain to obtain an additional 345 captains. Although the nondegree second lieutenants were less likely to reach captain by 2008, this group accounted for more than one-third of the 2001 cohort who

---

[3]   Lieutenants must also have completed BOLC B, a military education branch and basic soldiering course, before they are eligible for promotion to captain.

Table 3.5
Educational Attainment of Incoming Army National Guard Second
Lieutenants, 2001–2009

Education Level	2001	2002	2003	2004	2005	2006	2007	2008	2009
Total number of incoming 2LTs	1,425	1,961	2,399	2,196	2,036	1,763	1,822	2,138	2,415
Percentage at each education level									
Less than high school diploma	0	<1	<1	<1	<1	<1	<1	<1	0
High school diploma	41	28	2	3	2	2	3	3	2
Some college	1	11	35	34	30	31	31	26	23
Bachelor's degree	53	55	59	59	63	63	61	68	71
Advanced degree	5	5	4	5	5	4	4	3	4
Unknown	<1	0	<1	<1	<1	0	0	<1	0

SOURCE: DMDC RCCPDS.

Table 3.6
Army National Guard Second Lieutenant 2001 Degree Attainment Status
and Rate of Promotion to Captain by 2009

Degree Attainment Status	Incoming 2LT 2001 (N)	Made CPT by 2008	
		N	%
Lacked 4-year degree 2001	598	345	58
Had 4-year degree 2001	826	596	72
Total	1,424	941	66

SOURCE: DMDC RCCPDS.

made captain. Moreover, the proportion without a degree has shrunk since 2001; approximately one-quarter of 2009 OCS entrants lacked a four-year degree. Interviewed TAGs reported that these numbers are easy to address and that they create degree plans for those OCS graduates who lack four-year degrees. These soldiers are placed in the

16-month traditional, part-time, OCS programs, which give them more time to work on the degree than they would have in the accelerated OCS program. The Assistant Secretary of the Army for Manpower and Reserve Affairs (ASA[M&RA]) recently suspended removal provisions for first lieutenants who were twice nonselected for promotion to captain based on their lack of a baccalaureate degree. This policy allows officers additional time in the SELRES to complete civilian education prior to being removed.

We do, however, recommend requiring a four-year degree to commission as a second lieutenant. Soldiers could still enter OCS without a degree but would commit to completing it during the 16-month traditional OCS program. This provision would act to screen out soldiers with a low expectation of completing a four-year degree. Also, OCS commanders could be held accountable for ensuring that soldiers complete their degrees while they are in the program. As a result, commanders would have an explicit incentive to help soldiers complete their degree and to counsel those not expected to complete a degree not to seek entry in the OCS program.

If this policy were enacted, it would also affect those graduating via the ECP in the MJCs. The MJC graduates would no longer obtain an early commission after receiving their two-year MJC degrees but would be required to complete a four-year degree before receiving a commission. Although we support this recommendation, we recognize that the Army might choose to delay implementing it until the captain fill rate in the ARNG has reached satisfactory levels.

Other Sources of Commissioning

There are other current and pending sources of commission that we did not investigate. The federal OCS program is growing in terms of missions for FY 2010 and beyond, but it provides only small numbers of officers to the RCs (e.g., the expected FY 2010 distribution is 35 for the ARNG). ARCD also recently gained a mission to accession officers from the IRR. The IRR population has declined over the past three years, but ARCD is unsure whether that change reflects more soldiers staying in the AC or more joining the SELRES. Some interviewees suggested that ARCD also take on a mission of recruiting civilians with prior service. Others argued, however, that this mission would be exceedingly time-consuming to implement, with a potentially low return on investment. Finally, the ARNG is attempting to change legislation to allow it to commission from the Merchant Marine Academy, and it would like to gain newly commissioned officers from the U.S. Military Academy. Although these other sources can add to the supply of company-grade officers, their contribution seems likely to remain small (with the possible exception of IRR transfers) in comparison to ROTC, state OCS, and direct commission contributions.

The National Guard Academy Concept

There is also a proposal under consideration to create a new National Guard academy. H.R. 111-166 stipulates consideration of whether this concept would be a partial and feasible solution to the current company-grade officer shortfall.

The proposal under consideration is authored by the Missouri state adjutant general (Danner, 2010). It calls for the creation of a National Guard academy as a four-year, accredited military college that would supply fully degreed, high-quality commissioned officers to the ARNG. The education would combine both military and academic training, with an emphasis on National Guard federal and state missions. This proposal estimates that, by 2015, a new National Guard academy could enroll 250 cadets annually. These cadets could graduate in 2019 (assuming a four-year completion rate) and be promoted to captain by 2023 (assuming on-time promotion rates and current TIG regulations). By then, our modeling indicates, the ARNG could achieve a 100-percent captain fill rate through increasing accession and promotion rates, assuming that incentives affect accessions.

Although these accession practices are likely to have costs associated with them, these costs would not approximate those of establishing and continuing to operate a new postsecondary institution. Although we are not certain what the average cost per graduate of a new National Guard academy would be, if the caliber of the academy were similar to the U.S. Military Academy, the Naval Academy, and the Air Force Academy, a cost of $400,000 per graduate would not be out of line. In contrast, the costs of incentives, such as affiliation or retention bonuses or financial incentives for ROTC cadets to agree to serve in the SELRES rather than the IRR after they complete their ADSOs, are likely to be well below the cost per academy graduate. Therefore, since more-aggressive use of existing channels should eliminate the company-grade officer shortfall within the same time period and very likely at lower cost, we conclude that establishing a new National Guard academy is not a cost-effective solution to the company-grade officer shortfall.

However, establishing such an academy could have other educational or leadership benefits that we have not assessed. We therefore asked TAGs we interviewed whether they would support a new national academy. Their answers made it clear that they would need to be convinced that a new institution would meet their state's needs. These needs include

- mechanisms to attract from the national market, avoiding policies, advertising, and other factors that would appeal to a specific niche or region
- ensuring that an academy and student funding would not negatively affect ROTC; TAGs noted that ROTC allows commanders to select and groom their own lieutenants.

Finally, one TAG expressed concern that a new academy would be antithetical to such concepts as One Army, Total Force Integration, and Continuum of Service and warned the ARNG not to "make firewalls where we're trying to take them down."

CHAPTER FOUR

Promotion and Retention

In this chapter, we discuss options for changing promotion practices and improving retention rates. The captain fill rate could be improved by promoting lieutenants to captain more quickly or by retaining officers at the captain level for the maximum TIG. We also provide recommendations for improving retention rates, but these might be best considered as strategies to maintain already-high retention rates as the economy improves.

Promotion

A recent policy change has shortened the TIG from first lieutenant to captain in the RCs. Minimum TIG for promoting a second lieutenant to first lieutenant is 18 months, followed by a temporary reduction to a 24-month TIG requirement for first lieutenants before promotion to captain. In FY 2015, the minimum TIG from first lieutenant to captain reverts to 29 months. Now, in FY 2011, an officer can move from a second lieutenant to the rank of captain in as little as three years and five months.

Historically, lieutenants have not been promoted to captain after only four years of service, however. Indeed, data on second lieutenants newly commissioned in 2002 indicate longer time frames for promotion to captain. Seven years later, by 2009, 56 percent of these ARNG second lieutenants and 63 percent of these USAR second lieutenants had been promoted to captain. In the ARNG, of those who were promoted in seven years, most (56 percent) were promoted with five to six years TIG as second and first lieutenants, but a significant proportion (38 percent) was promoted with six to seven years TIG as lieutenants. In the USAR, of those who were promoted within seven years, about half (48 percent) were promoted with five to six years TIG as lieutenants, and 46 percent were promoted with six to seven years TIG as lieutenants.

It is possible that, of those who were promoted, a higher proportion of ARNG officers were promoted sooner because of the ARNG's use of unit vacancy boards (UVBs). In the ARNG, when there is a vacancy in a unit, officers eligible for the position can be put up for promotion with less TIG than is required for the regular DA promotion board. The USAR has a similar system, PVBs, but, according to interviewees, it has not used these boards for several years.

In general, promotion to captain might be slower than the minimum TIG requirements because of the need to promote reserve officers within the region where they work and live. In some states and regions (and for some occupational specialties), vacancies might be available more frequently than in others.

It is also possible that frequent cross-leveling delays promotion, as noted by Feidler (2008). He noted that an officer cross-leveled to a new unit for deployment is not eligible for promotion until he or she returns to his or her home unit. Thus, deployment can contribute to delayed promotion.

Some interviewees expressed concern about getting officers into BOLC B[1] in time for promotion to captain. Second lieutenants are expected to complete BOLC B within 18 months of commissioning, although they could be granted extensions. All lieutenants must have completed BOLC B before they are promotable to captain. Table 4.1 presents Army G-1 data on average BOLC B wait times by component and branch. It compares average wait times for active-duty officers (shown in the first column) with those for newly commissioned lieutenants from ROTC in both the USAR and the ARNG and from ARNG OCS programs. The AC officers are also coming out of ROTC.

---

[1] The length of BOLC B ranges from about ten to 40 weeks, depending on the officer's branch.

Table 4.1
Number of Days Newly Commissioned Second Lieutenants in 2009 Waited
Before Starting Basic Officer Leaders Course Part B, by Component and
Branch

Branch	AC (ROTC)	USAR (ROTC)	ARNG (ROTC)	ARNG OCS
FI	42	381	293	Does not track by branch but estimates average of 9–10 months overall.
QM	60	256	159	
OD	88	266	174	
TC	94	226	182	
FA	108	Not applicable	134	
MI	112	299	187	
AD	113	Not applicable	148	
MS	116	337	178	
SC	116	246	172	
EN	118	237	199	
AG	127	335	243	
CM	135	284	145	
MP	150	249	150	
AR	179	340	182	
IN	189	300	216	
AV	219	279	310	
Average	135.6	287	192	285

SOURCE: Army G-1.

NOTE: Dark green indicates less than 150 days. Light green indicates 151–200 days.
Yellow indicates 201–300 days. Red indicates more than 300 days.

Although wait times vary by component and branch, it appears
that, on average, lieutenants are getting a seat prior to the 18-month
deadline. Because seat availability varies by branch, interviewees
acknowledged that some lieutenants have had to switch their selected
branch to get through BOLC B on time. The wait time for BOLC B
is longer in the USAR than in either the ARNG or the AC, and short-

est in the AC. However, interviewees in the USAR reported that the vast majority of lieutenants progress through BOLC B in time to be promoted to captain.

Despite an inability to pinpoint the barriers to promotion, our data indicate that few serve as a lieutenant for only four years. Furthermore, our data indicate that, on average, captains are promoted to major after only five years (two years short of the maximum TIG).

Retention

Literature, interview results, and secondary data analyses indicate that RC officer continuation rates have held steady or improved since 2006. We looked at loss rates by component from one fiscal year to the next as a proportion of the officer population. From 2008 to 2009, 6 percent of company-grade officers left the ARNG and 8 percent of company-grade officers left the USAR. Table 4.2 presents the percentage of captains who left the component in each year from 2001 through 2009. We present captains in this table because officers typically cannot voluntarily leave RC service until they have fulfilled an eight-year MSO. Although there are losses of lieutenants, a higher proportion of captains leave in any given year.

As the table demonstrates, captain loss rates have held steady in the ARNG and improved slightly in the USAR in the past ten years. Interviewees agreed that current retention rates are not problematic.

Table 4.2
Annual Loss Rates of Captains in the Army National Guard and U.S. Army Reserve, 2001–2009

	Percentage of Total per FY								
	2001	2002	2003	2004	2005	2006	2007	2008	2009
ARNG	8	7	7	8	8	7	8	8	8
USAR	13	13	11	12	12	9	10	9	10

SOURCE: DMDC RCCPDS.

Nonetheless, some speculated that attrition could increase when the economy improves.

Further research on retention incentives might be needed to help the components address a potential increase in attrition. Bonuses have been shown to be effective and cost-effective in increasing recruiting and retention in the AC (e.g., see Mattock and Arkes, 2007, and Asch et al., 2010) and effective in attracting to the RC enlisted AC members who do not reenlist to the AC (Hosek and Miller, 2010) but have not yet been studied with respect to the recruiting and retention of RC officers. Survey data, personnel data, and pay data would be helpful in studying the effect of special and incentive pays, as well as of command climate, deployment, and other factors on RC officer retention.

An initiative worth monitoring is a pilot program implemented by the National Guard Bureau (NGB) in Kentucky that allows local authority over allocation of incentive dollars for the ARNG enlisted population. Under this program, units identify their needs and can target financial incentives for enlistment and retention accordingly. State TAG staff members work with units down to the battalion level to identify needs by paragraph and line. Units determine whether there is a need, for example, to fill a particular MOS or grade, or to retain a particular soldier with a specific MOS and certain number of years of experience. If successful, this program could be implemented across the states in FY 2011. Indeed, the ARNG would like more local control of incentive dollars. Further study could examine trade-offs between local control and equity.

CHAPTER FIVE

Conclusions and Recommendations

In this chapter, we present conclusions and recommendations related to the captain shortfall. Although our recommendations stem from our analysis of Army data, we hope that they are relevant for the USMC and U.S. Navy RCs as well. We also provide suggestions for further study.

Fill Rates

Although the overall company-grade officer fill rates in Army RC units have been improving gradually due to slowly increasing captain fill rates and lieutenant fill rates that increasingly exceed 100 percent, aggressive measures would be needed to dramatically improve the captain fill rate in the immediate future. The ARNG has steadily increased its inventory of lieutenants during the past seven years. The number of lieutenants in the USAR has fluctuated more over time, with a steady increase in the past three years. In both RCs, the number of captains has increased during the past five years. However, these increases have been small and have not allowed either RC to exceed a 75-percent captain fill rate in the past seven years.

Although this monograph focuses on company-grade officers, we observed that there is also a shortfall in the inventory of majors, compared to authorizations, in both Army RCs. In FY 2009, the major fill rate in the ARNG was 74 percent; it was 89 percent in the USAR. The gap between major authorizations and inventory appears to be long-standing, particularly in the ARNG. In the USAR, the major fill rate has declined by about 30 percentage points over seven years. The number of majors in each RC has decreased over the past seven years while the authorizations have increased in the ARNG and remained fairly stable in the USAR. The major shortfall is worsening as the captain shortfall is improving, albeit moderately.

Projections

Our modeling demonstrates that the Army RCs could get to a 100-percent captain fill rate in five to ten years.

The most-aggressive model results in a 100-percent captain fill rate in the ARNG in 6.5 years, and in five years in the USAR. In this model, we assume that accessions increase by 10–30 percent over the next three years, followed by a leveling-off period. This modeling assumes historical average continuation and promotion rates (if lieutenants were promoted to captain more quickly, the captain fill rate could reach 100 percent even earlier). Accessioning 10 percent more officers in 2010, followed by 20 percent more than the 2009 numbers the following year and 30 percent more the next, would entail aggressive accessioning policies and practices, but it could be worth investing in such policies and practices in the short term, particularly given the current unemployment context, to dramatically boost the captain fill rate.

The most-conservative model results in a 100-percent captain fill rate in the ARNG in 10.5 years, and in nine years in the USAR. In this model, accessions increase by 4 percent in the ARNG and 6 percent in the USAR each year for the next 10.5 years. Recent average loss rates continue. Lieutenants are promoted to captain after serving as a second and then a first lieutenant for a total of four years. On average, captains are promoted to major after five years, which is the current average TIG.

Policy Recommendations

Although we present policy options here to improve accession, retention, and promotion rates, we do note that it could also be important to reexamine the authorization structure in the RCs. Although gaps between inventory and authorizations seem quite large, responses to our interview questions on the impact of the captain shortfall lead us to question the extent to which the shortfall is problematic. It could be that, because the shortfall has been in existence for so long, commanders have become accustomed to implementing strategies to address the problem and, therefore, no longer view it as significant. If that is the case, it could be worth investigating the extent to which the current authorizations are warranted. The U.S. Army Chief of Staff has directed TRADOC to review authorizations for majors and lieutenant colonels. This review could provide an opportunity to subsequently review captain authorizations.

As noted above, however, even if the authorization structure changes, it is likely that both components will need more captains (and majors) than they are now routinely accessioning. Our modeling demonstrates that, by increasing accessions slightly, the captain fill rate would reach 100 percent in nine to ten years. More-dramatic increases would fill the captain requirements in five years or so. Either approach is likely to be more feasible in the current economic context than might be the case in the future.

Active Component–to–Reserve Component Transfers

We recommend investing in increasing the AC-to-RC transfer rate. The reported obstacles to increasing this rate appear surmountable, and this accession route is most likely to result in an immediate surge of new captains.

There are several options that could increase the number of AC officers transferring to an RC that include improving data accuracy, counseling, and the transition process in general. Improving counseling would necessitate gathering better data on AC officers nearing separation and on RC officer vacancies. One option to improve the transfer process is to change the statute to allow a single commission. Another option is to allow an exiting AC officer to sign the RC commission at the time of exit and wait to date the commission until the AC resignation is approved. A third option is to offer exiting AC officers the opportunity to sign a letter of intent to join an RC and selectively offer an increased bonus (depending on the extent of shortage and the criticality of the officer's specialty) as an incentive to sign. Other incentives might include graduate school reimbursement or job opportunities with federal agencies. In making decisions about incentives, it will be important to predict the effect that they will have on the AC officer pool.

Reserve Officer Training Corps and the Officer Direct Commission Program

ROTC is likely to continue to be an important source of commissioning for both RCs. The recent policy change that requires each component to achieve its mission before any component can overrecruit should ensure that the RCs get their "fair share" of ROTC cadets. However, we recommend that policymakers consider investments in incentives for cadets to join the RCs to reduce the number commissioned from the bottom of the OML. At the time of contracting, cadets need information about the RCs so that they can make an informed choice about components and the scholarships each can offer. Cadets could also be offered a range of incentives to join an RC, including

- bonus
- choice of branch (recognizing that this might not be feasible for all branches, given regional variation)
- guaranteed internships
- veteran's preference rights for federal government civil service positions to those on DEDNG scholarships
- simultaneous eligibility for DEDNG scholarship students for the MGIB-SR
- graduate school funding.

Many interviewees support the notion of changing ROTC contracts so that cadets who receive initial AC commissions are required to fulfill their MSOs in a drilling unit rather than in the IRR. An alternative to requiring such service would be to structure the initial contract such that AC officers who have finished their ADSOs are automatically transferred into the SELRES, forcing them to proactively "opt out" if they want to join the IRR instead. Another related option, again at the point of contracting, would be to offer incentives to AC officers to commit to serve time in the SELRES after their ADSOs rather than in the IRR.

There are already proposals under consideration to directly commission more individuals with specific training or skills as captains in the Army RCs. There are concerns, however, about the ability of newly commissioned captains to lead troops. It should be possible to utilize direct commissions more extensively for occupations that draw on civilian skills for initial assignments within staff organizations, accompanied by expanded leadership education and training to prepare these officers for subsequent assignments within operational units.

Army National Guard State Officer Candidate School Programs

H.R. 111-166 requested the Secretary of Defense to examine whether soldiers entering state OCS programs should be required to possess a baccalaureate degree. This question stems from the concern that many officers who graduate from a state OCS program never complete their baccalaureate degree and, therefore, are never promoted to captain. Approximately 42 percent of the lieutenants commissioned via OCS without a four-year degree in 2001 had left the service by 2008.[1] Fifty-eight percent had been promoted to captain (and had therefore obtained their degree) and were retained through 2008. This proportion is less than the 72 percent of those who entered OCS with a four-year degree and were promoted to captain and retained in service through 2008.

Despite this statistically significant difference, we recommend continuing the practice of allowing soldiers with 60 or more credit hours to enroll in an OCS program. Policy decisions need to be made within the context of a large shortfall in captain inventory. Most of

---

[1] Forty-one percent is the proportion that left the service by 2009. It is possible that some of these officers had obtained a four-year degree prior to leaving.

the OCS soldiers who lacked four-year degrees in 2001 did become captains, and, of all second lieutenants who were promoted to captain by 2008, one-third started without a degree. Individual TAGs and the NGB have put several programs in place to support these soldiers lacking a four-year degree. And the proportion entering state OCS programs without a four-year degree is shrinking.

We do recommend, however, that OCS candidates complete their four-year degrees while they are going through OCS and that degree completion be required for an RC commission. This goal is shared by the TAGs we interviewed and is reportedly achievable if candidates go through traditional, rather than accelerated, OCS programs. If OCS candidates can complete their degrees during OCS, they will be ready and deployable assets to their components, and degree completion will not be stymied by deployments, as is the case now. If this policy is implemented, it will necessitate changes to the ECPs at the MJCs as well. Although we support this recommendation, we recognize that the Army might not want to implement it until the captain fill rate reaches a satisfactory percentage.

National Guard Academy

H.R. 111-166 also requested the Secretary of Defense to consider whether establishing a new National Guard academy would be a partial and feasible solution to the company-grade officer shortfall. To address this question, we relied on published projections that a new National Guard academy could enroll 250 cadets annually, starting in 2015. We assume that all of these cadets graduate in 2019 and are promoted to captain by 2023. By then, our modeling indicates, the ARNG could have achieved a 100-percent captain fill rate through aggressive, but achievable, accessioning and strategic promotion practices, including offering incentives that are effective in shaping individuals' accession and retention behavior.

Although these accession practices are likely to have costs associated with them, these costs would not likely approximate those of establishing and continuing to operate a new postsecondary institution. Therefore, although we acknowledge that a new academy is a feasible source of new captains, it would not eliminate the shortage any sooner than would other methods and would very likely cost more. Therefore, we conclude that establishing a new National Guard academy is not a cost-effective solution to the company-grade officer shortfall.

However, establishing such an academy might have other educational or leadership benefits that we have not assessed. It might address problems in the ARNG other than the company-grade officer shortfall and could therefore be an effective solution to problems not evaluated in this monograph.

Promotion and Retention

We recommend promoting lieutenants to captain more quickly than is now the case, but not in less time than is mandated by law. To do so, the USAR might need to accelerate the use of vacancy boards. Other barriers might have to be examined as well, including regional variation in opportunities and the impact that cross-leveling has on promotion rates.

Because the literature, our data analysis, and our interviewees concluded that current retention rates are relatively high compared with historical rates, we did not manipulate loss rates in our modeling. However, the ARNG and USAR could also benefit from improved retention rates or need to maintain current high retention rates as the economy improves. Incentives, such as bonuses or support for graduate education, could be effective policy tools for increasing retention. The demonstration program in Kentucky that allows units to make decisions on how to spend incentive funding to affect both enlistments and retention should be carefully studied and, if successful, considered for officer populations in both Army RCs.

Further Analysis

Findings in this monograph could benefit from further analysis. Indeed, throughout the monograph, we have provided recommendations for further study. For example, we suggest further study on retention incentives. In this section, we propose further study on four key topics in this monograph: the structure of lieutenant and captain authorizations, AC-to-RC transfers, promotion practices, and ARNG training.

In this monograph, we have focused on increasing the supply of captains; we also recommend that the Army conduct an analysis of its force structure. We recommend this analysis because our interview and secondary data analyses were unsuccessful in identifying important problems emanating from the captain shortfall. Key questions that could be addressed in such an analysis include the following:

1. What captain positions are being filled by lieutenants or are being left vacant?
2. Could these positions be recoded to a lower rank?
3. For those positions filled by lieutenants, is there a cascading effect that leaves lieutenant positions vacant?
4. What is the impact of cross-leveling officers across positions requiring different ranks?

Understanding the specific requirements of the vacant positions could provide additional guidance on reclassifying positions, direct commissioning (lateral entry) at higher ranks, or even eliminating the authorization.

We argue above that there could be room for increasing the AC-to-RC transfer rate. Further analysis could explore AC officers' retention behavior and motivation to join the Army RCs after they leave AC service and to build demonstration projects to test promising approaches to increase AC-to-RC transfers. Data from interviews with a sample including both exiting and staying officers could be analyzed to better understand their decisionmaking processes. These data could be used to develop approaches (e.g., better counseling, different incentives, sabbaticals between AC and RC service) to increase the proportion that transfers to an RC. For example, one approach could be to better connect job opportunities with service in an RC.

In terms of promotion practices, further analysis could facilitate the RCs' ability to promote lieutenants to captain at minimum TIG and accelerate the use of vacancy boards, particularly in the USAR. It would be important to understand the barriers to vacancy promotions (e.g., regional), as well as the limits of this system. It would also be important to understand the impact that cross-leveling could have on promotion in both RCs.

Although this study concluded that a National Guard academy is not a cost-effective solution to the company-grade officer shortfall, it might be worth studying whether and how ARNG officer education could be improved. Further study could focus on curriculum design, investigating the content of current curriculum, and ascertaining whether additional or different emphases could benefit the dual mission of the ARNG. The current proposal to establish a new academy has reportedly generated excitement among several constituencies, which might indicate a need for new ARNG education models.

Unit Type Code Definition Table

Table A.1 provides the names of the abbreviated unit types shown in Tables 2.1 and 4.1.

Table A.1
Unit Type Codes

Unit Type Code	ARNG Unit Type	USAR Unit Type
AD	Air defense artillery	Not applicable
AG	Adjutant general	Adjutant general
AQ	Contingency contract support battalion	Contingency contract support battalion
AR	Armor	Armor
AV	Aviation	Aviation
CA	Not applicable	Civil affairs
CH	Chaplain	Chaplain
CM	Chemical	Chemical
CS	Composite service	Composite service
EN	Engineer	Engineer
FA	Field artillery	Field artillery
FI	Financial management	Financial management
HQ	Headquarters	Headquarters
IN	Infantry	Infantry

Table A.1—Continued

Unit Type Code	ARNG Unit Type	USAR Unit Type
IO	Information operations	Information operations
JA	Not applicable	Judge advocate
MD	Medical corps	Medical corps
ME	Mechanized	Mechanized
MI	Military intelligence	Military intelligence
MP	Military police	Military police
MS	Medical support	Medical support
OD	Ordnance	Ordnance
PI	Public affairs	Public affairs
PO	Not applicable	Psychological operations
QM	Quartermaster	Quartermaster
SB	Reconnaissance and support battalions	Not applicable
SC	Signal corps	Signal corps
SF	Special forces	Special forces
SP	Medical specialist corps	Medical specialist corps
TC	Transportation	Transportation
TDA	Table of distribution and allowances	Table of distribution and allowances

References

AR 600-8-29—*See* Headquarters, Department of the Army, 2005.

Asch, Beth J., Paul Heaton, James Hosek, Francisco Martorell, Curtis Simon, and John T. Warner, *Cash Incentives and Military Enlistment, Attrition, and Reenlistment*, Santa Monica, Calif.: RAND Corporation, MG-950-OSD, 2010. As of January 29, 2011:
http://www.rand.org/pubs/monographs/MG950.html

Bonn, Keith E., *Army Officer's Guide*, Mechanicsburg, Pa.: Stackpole, 2005.

Danner, Stephen L., "Creating and Establishing a National Guard Service Academy," paper, Office of the Adjutant General, Missouri National Guard, 2010.

Defense Science Board, Task Force on Deployment of Members of the National Guard and Reserve in the Global War on Terrorism, Office of the Under Secretary of Defense for Acquisition, Technology, and Logistics, *Defense Science Board Task Force on Deployment of Members of the National Guard and Reserve in the Global War on Terrorism*, Washington, D.C.: Office of the Under Secretary of Defense for Acquisition, Technology, and Logistics, September 2007. As of June 14, 2010:
http://www.acq.osd.mil/dsb/reports/ADA478163.pdf

Erlandson, Ernest A., *An Institution in Crisis: The Army Reserve Officer Corps*, Carlisle Barracks, Pa.: U.S. Army War College, February 16, 2009. As of January 29, 2011:
http://handle.dtic.mil/100.2/ADA498034

Feidler, Robert, *Report on the Junior Officer Shortage Program*, Defense Education Forum of the Reserve Officers Association of the United States, December 11, 2008. As of June 14, 2010:
http://www.roa.org/site/DocServer/
JO_SHORTAGE_rpt_11dec.pdf?docID=14681

GAO—*See* U.S. General Accounting Office.

Headquarters, Department of the Army, *Officer Promotions*, Army Regulation 600-8-29, February 25, 2005. As of February 3, 2011:
http://armypubs.army.mil/epubs/pdf/R600_8_29.pdf

———, "Army Demographics: FY09 Army Profile," September 30, 2009. As of February 3, 2011:
http://www.2k.army.mil/downloads/FY09%20Army%20Profile.pdf

Hosek, James, and Trey Miller, *Effects of Reserve Enlistment and Affiliation Bonuses on Prior Service Enlistment in the Selected Reserve*, Santa Monica, Calif.: RAND Corporation, unpublished research, 2010.

Howe, Richard D., *Mid-Grade Army Reserve Officers: In Short Supply of a Critical Component of a Strategic Means*, Carlisle Barracks, Pa.: U.S. Army War College, March 2005. As of January 29, 2011:
http://handle.dtic.mil/100.2/ADA433671

H.R. 111-166—*See* U.S. House of Representatives, 2009.

Mattock, Michael, and Jeremy Arkes, *The Dynamic Retention Model for Air Force Officers: New Estimates and Policy Simulations of the Aviator Continuation Pay Program*, Santa Monica, Calif.: RAND Corporation, TR-470-AF, 2007. As of January 29, 2011:
http://www.rand.org/pubs/technical_reports/TR470.html

Mattock, Michael G., Beth J. Asch, James R. Hosek, Christopher Whaley, and Christina Panis, *Modeling Options for Improving Mid-Career Officer Retention*, Santa Monica, Calif.: RAND Corporation, unpublished research.

Public Law 103-337, National Defense Authorization Act for Fiscal Year 1995, October 5, 1994. As of February 3, 2011:
http://thomas.loc.gov/cgi-bin/bdquery/z?d103:s.02182:

Public Law 111-84, National Defense Authorization Act for Fiscal Year 2010, October 28, 2009.

Reno, LTG (ret.) Bill, then Army Deputy Chief of Staff for Personnel, "Resolution of Officer Issues in the United States Army (The Reno Report)," memorandum for the Chief of Staff of the Army, 2008.

Reserve Officer Training Corps, U.S. Army Cadet Command, Army National Guard, "Army National Guard (ARNG) Scholarship Information," undated web page. As of February 7, 2011:
http://www.rotc.usaac.army.mil/command/ng/ng_GRFDscholarships.html

U.S. Code, Title 10, Armed forces, Subtitle A, General military law, Part II, Personnel, Chapter 37, General service requirements, Section 651, Members: required service. As of February 3, 2011:
http://frwebgate.access.gpo.gov/cgi-bin/usc.cgi?ACTION=RETRIEVE&FILE=$$xa$$busc10.wais&start=2308962&SIZE=10684&TYPE=PDF

———, Title 10, Armed forces, Subtitle E, Reserve Components, Part II, Personnel generally, Chapter 1201, Authorized strengths and distribution in grade.

————, Title 10, Armed forces, Subtitle E, Reserve Components, Part IV, Training for Reserve Components and educational assistance programs, Chapter 1606, Educational assistance for members of the selected reserve.

U.S. General Accounting Office, *Reserve Officers' Training Corps: Less Need for Officers Provides Opportunity for Significant Savings: Report to the Chairman, Subcommittee on Military Personnel and Compensation, Committee on Armed Services, House of Representatives*, Washington, D.C., GAO/NSIAD-91-102, May 1991. As of December 15, 2010:
http://archive.gao.gov/d20t9/143856.pdf

U.S. House of Representatives, *National Defense Authorization Act for Fiscal Year 2010, June 18, 2009*, 111-1 House Report 111-166, Washington, D.C., 2009. As of January 29, 2011:
http://www.gpo.gov/fdsys/pkg/CRPT-111hrpt166/pdf/CRPT-111hrpt166.pdf

Whitlock, Joseph E., *Can the Army Reserve Overcome Its Growing Company Grade Officer Shortage?* Fort Leavenworth, Kan.: School of Advanced Military Studies, U.S. Army Command and General Staff College, AY 01-02, 2002. As of January 29, 2011:
http://cgsc.contentdm.oclc.org/u?/p4013coll3,241

————, *How to Make the Army Force Generation Work for the Army's Reserve Components*, Carlisle, Pa.: Strategic Studies Institute, U.S. Army War College, August 2006. As of January 29, 2011:
http://purl.access.gpo.gov/GPO/LPS77740